D1543417

THE GRAVITATIONAL FORCE PERPENDICULAR TO THE GALACTIC PLANE

A Meeting held at the Center for Galactic Astronomy,
Western Connecticut State University in Danbury, Connecticut
May 19 - 20, 1989

Sponsored by Perkin-Elmer Corporation, The University Foundation
of Western Connecticut State University and Van Vleck Observatory.

Edited by

A. G. Davis Philip
Van Vleck Observatory and
Union College

Phillip K. Lu
Center for Galactic Astronomy,
Western Connecticut State University

Van Vleck Observatory Contribution No. 9

L. Davis Press
Schenectady, New York
1989

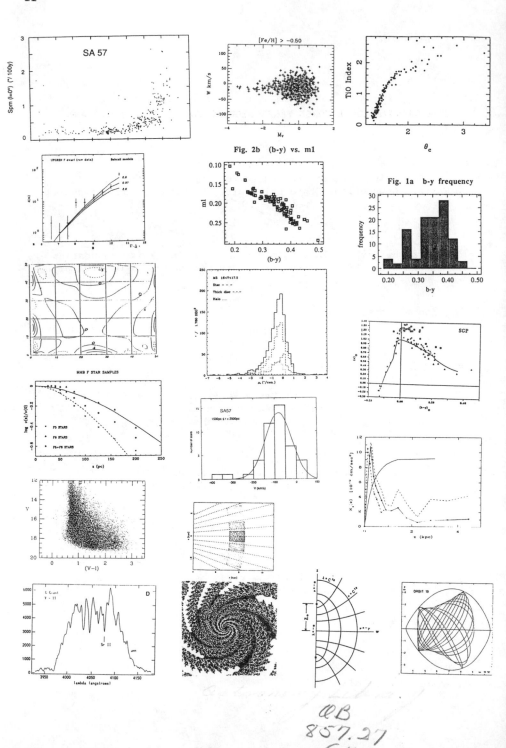

TABLE OF CONTENTS

Oral Papers

Poster Papers

Indices

PREFACE

The meeting on the <u>Gravitational</u> <u>Force</u> <u>Perpendicular</u> <u>to</u> <u>the</u> <u>Galactic</u> <u>Plane</u> was held May 19 - 20, 1989 at Western Connecticut State University, Danbury, Ct. The idea to hold this meeting was generated about a year ago by a small group at the Van Vleck Observatory during the meeting on <u>Calibration</u> <u>of</u> <u>Stellar</u> <u>Ages</u>. Although our original concept was much smaller in scope, involving mostly local astronomers, the workshop received enthusiastic support from many people who attended the 20th IAU General Assembly in Baltimore, Maryland in August, 1988. An announcement of an expanded program appeared in IAU TODAY on August 13, 1988.

Much work has been done on missing mass and dark matter, notably by John Bahcall and his colleagues at the Institute for Advanced Studies, Kenneth Freeman of the Mount Stromlo Observatory, Gerry Gilmore of Cambridge, England, Ken Yoss of the University of Illinois and Paul Schechter of MIT. Many people suggested that a meeting of this subject would be an excellent way to exchange information as to who is doing what. The title of the meeting, <u>The</u> <u>Gravitational</u> <u>Force</u> <u>Perpendicular</u> <u>to</u> <u>the</u> <u>Galactic</u> <u>Plane</u>, was suggested by Ivan King while I was on sabbatical leave at CTIO during the Fall of 1988.

Members of The Scientific Organizing Committee were:

Phillip K. Lu	W. Conn. St. Univ (Chairman)
John Bahcall	Inst. for Advanced Study
Kenneth Freeman	Mt. Stromlo Observatory
Gerald Gilmore	Institute of Astronomy
Ivan King	Univ. of California, Berkeley
A. G. Davis Philip	Van Vleck Obs. and Union College
Arthur R. Upgren	Van Vleck Obs.

The Local Organizing Committee included:

Dennis Dawson   (Chairman)
Christopher Ftaclas
Francis Kendziorski
William van Altena

Financial support was provided by:

The Perkin-Elmer Corporation
The University Foundation
Western Connecticut State University
Van Vleck Observatory

Thirty-nine participants from nine countries attended the meeting.

The Chairmen of the sessions were:

       I. P. K. Lu
      II. P. Schechter
    III. K. Freeman
    IV. G. Gilmore
     V. A. G. D. Philip
    VI. W. van Altena
   VII. A. R. Upgren
  VIII. I. King

I would like to thank Dr. Stephen Feldman, President of Western Connecticut State University, Mr. John Rehnberg, Vice President of Space Sciences Division of Perkin-Elmer Corporation and Dr. Arthur Upgren, Director of the Van Vleck Observatory for their support, which covered all costs of printing and publishing the Proceedings of the conference. Thanks are also due to Dr. Philip Steinkrauss, Vice President for Academic Affairs and Dr. Carol Hawkes, Dean of the School of Arts and Sciences, who provided funds and hosted luncheons for the meeting. I would like to a give special thanks to Dr. Dallas Beal, President of the Connecticut State University, who came from New Britain to host the symposium banquet.

A number of departments and divisions within the university helped prepare for the meeting, including the Media Center, Public Affairs, the University Police and maintenance personnel. Much gratitude goes to Diane Golden, Director of University Design and Production, Ruth Corbett, Director of Research and Grants, John Wallace, Director of Housing, Dave Smith and Heidi Upton of the Music Department and Linda Cuffee and Irene Duffy who managed and prepared the gathering and typing of the questions and answers during the meeting. William Quinnell took the group photograph.

I would also like to thank my assistants and students. Jeff Miller deserves special thanks for his leadership in organizing the computer files and arranging for projection of slides. Those students who worked during the meeting were Donald Platt, Joel Gomes, Stephen Veillette, Mike Bova and Valois Costa. Finally, I would like to thank my wife Catherina, who supported the meeting in many ways and who hosted the reception at our house.

                           Phillip K. Lu

Danbury
July 29, 1989

Electronic methods of communication have made the production of
conference proceedings much more rapid and have increased the accuracy
with which the papers can be prepared.   Authors of papers at the K_z
meeting sent their papers to Schenectady either by BITNET or on a
floppy disk which allowed them to be transferred to the word processor
Final Word.   Non-ASCII characters were replaced, using the paper copies
of each contribution sent by mail.   The LaserJet output was then sent
back to authors for proofreading and corrections were returned to
Schenectady for production of the final version.   I would like to thank
authors for sending in their papers promptly after the meeting and for
communicating their corrections to me.   It is this cooperative effort,
between authors and the editors, that allows the rapid, accurate
publication of the proceedings.

The figure on the cover of the book is from the thesis of William
I. Hartkopf at the University of Illinois, A Kinematic and Abundance
Survey of the Galactic Pole Regions and presents a number of K_z
solutions.   Additional plots may be found in the paper by Ken Yoss.
The letters in the figure refer to solutions made by the following
investigators:

    a: Oort, J. H.   1932 B. A. N. 6, 249
    b: Hill, E. R.   1960 B. A. N. 55, 1
    c: Oort, J. H.   1960 B. A. N. 55, 45
    d: Upgren, A. R. 1962 Astron. J. 67, 37.
    e: Perry, C. L. 1969 Astron. J. 74, 139. (Case I)
    f: Perry, C. L. 1969 Astron. J. 74, 139. (Case II)
    g: Hartkopf, W. I. 1981 Ph.D. Dissertation, Univ. of Illinois.

During the meeting suggestions were made that this figure should appear
on the cover of the proceedings and Ken Yoss kindly provided a copy of
the figure for this purpose.

Kristina Philip is thanked for typing some of the material into
the word processor.   Dennis Dawson is thanked for proofreading the
manuscript before publication.

                              A. G. Davis Philip

Schenectady
07/30/89

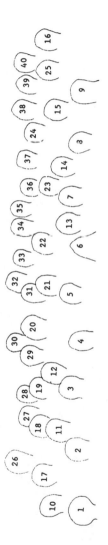

Identification of People in Conference Photo

1. Linda Cuffee
2. Hong-Yee Chiu
3. Dennis Dawson
4. Noreen Grice
5. Ken Croswell
6. Annie Robin
7. Chris Flynn
8. Kenneth Yoss
9. Kavan Ratnatunga
10. Frank Kendziorski
11. Phillip Lu
12. Stephen Maran
13. Kenneth Freeman
14. Ivan King
15. John Norris

16. Davis Philip
17. Irene Duffy
18. William van Altena
19. Michel Crézé
20. O. F. Bienaymé
21. Birgitta Nordström
22. Thomas Statler
23. Gary Da Costa
24. Gerard Gilmore
25. Mike Bova
26. John Lee
27. Terry Girard
28. Ting-Gao Yang
29. Carlos Lopez
30. Jurgen Stock

31. Arthur Upgren
32. Robert Wing
33. Valois Costa
34. Paul Schechter
35. Jeffrey Miller
36. Andrew Gould
37. Charles Trefzger
38. Andreas Spaenhauer
39. Stephen Veillette
40. Don Platt

LIST OF PARTICIPANTS

ARGENTINA

Carlos Lopez   University of San Juan and Yale Univ. Obs.

AUSTRALIA

Kenneth Freeman   Mt. Stromlo Observatory
Christopher Flynn   Mt. Stromlo Observatory
John Norris   Mt. Stromlo Observatory

CHINA

Ting-Gao Yang   Sian Observatory and Yale Univ. Obs.

DENMARK

Birgitta Nordström   Copenhagen Observatory

FRANCE

O. F. Bienaymé   Besançon Observatory
Michel Crézé   Besançon Observatory
Annie Robin   Besançon Observatory

SWITZERLAND

Andreas Spaenhauer   Astronomy Institute, Basel
Charles Trefzger   Astronomy Institute, Basel

USA

Alan Adler   WCSU
Hong-Yee Chiu   Goddard Space Flight Center, NASA
Ken Croswell   Center for Astrophysics
Gary Da Costa   Yale University Observatory
Dennis Dawson   WCSU
Cristopher Ftaclas   Perkin Elmer
Terry Girard   Yale University Observatory
Andrew Gould   Princeton University
Noreen Grice   Boston Museum of Science
Frank Kendziorski   WCSU
Ivan King   University of California, Berkeley
John Lee   Yale University Observatory
Phillip Lu   WCSU
Stephen Maran   Goddard Space Flight Center, NASA
Jeffrey Miller   WCSU
Davis Philip   Union College and Van Vleck Obs.
Kavan Ratnatunga   Goddard Space Flight Center, NASA
Jim Rose   University of North Carolina
Paul Schechter   Massachusetts Institute of Technology
Thomas Statler   JILA, University of Colorado
Arthur Upgren   Van Vleck Observatory
William van Altena   Yale University Observatory
Edward Weis   Van Vleck Observatory
Robert Wing   Ohio State University
Kenneth Yoss   University of Illinois
Marie -C. Zolcinski   WCSU

UNITED KINGDOM

Gerard Gilmore   Institute for Astronomy, Cambridge

VENEZUELA

Jurgen Stock   CIDA, Merida

REVIEW PAPERS

Chairmen of Sessions

Session I	P. K. Lu
Session II	P. Schechter
Session III	K. Freeman
Session IV	G. Gilmore
Session V	D. Philip
Session VI	W. van Altena
Session VII	A. R. Upgren
Session VIII	I. King

COMPARED EFFICIENCY OF GLOBAL MODELING TIED TO GENERAL STAR COUNTS AND SELECTED SAMPLES AS K_z TRACERS.

Michel Crézé, Annie C. Robin and Olivier Bienaymé

Observatoire de Besançon

ABSTRACT: Early attempts to derive the K_z law from tracer samples tried to tackle the numerical solution of an ill-conditioned inverse problem followed by the computation of derivatives of this solution. Most recent investigations prefer to hypothesize different mass distributions, compute the behavior of tracer populations in the resulting potential and then choose the best hypothesis by comparing observed and predicted distributions of tracer stars.   Investigations based on a compromise between both approaches are likely to induce overconfidence in the reliability of tracer samples.   Recent claims for very large amounts of unseen mass in the galactic disk are based on such overconfidence.   We have made one further step in the direction of global modeling.   We gathered in a galaxy model any reliable a priori knowledge about stellar populations relevant to the problem, leaving only a few hypotheses to be tested against observed star counts.   Under each different hypothesis we did impose dynamical consistency.   As a consequence, we have been able to predict the behavior in each hypothesized potential of the whole disk population.   Hence general star counts can be used as observational constraints instead of spectral type selected samples in which homogeneity is not easy to assess for large samples at faint magnitudes.   The resulting dynamical constraint on the local density is strengthened by one order of magnitude with respect to former ones, even taking into account model uncertainties.   Then the dynamically estimated local mass density is shown to be compatible with the observable one within narrow error bars.   Although a dark matter corona is still imposed by rotation curve considerations, it contributes little to the local density and there is no need for any specific kind of dark matter in the disk.   This result, published in Bienaymé, Robin and Crézé (1987) is now confirmed by the analysis of Kuijken and Gilmore (1989) based on different grounds and fully independent data.

1.   INTRODUCTION

The basic tools in this K_z story are the Poisson equation and the collisionless Boltzmann equation.   The Poisson equation is the way towards   the   potential   once   you   know   the   total   mass   density

A. G. Davis Philip and P. K. Lu (eds.)
The Gravitational Force Perpendicular to
the Galactic Plane   3 - 18
©   1989   L. Davis Press

distribution. It is usually written:

$$\Delta\Phi = 4\pi G\rho \qquad (1)$$

Where Φ and ρ are functions of the three space coordinates and represent, respectively, the gravitational potential and the total volume density that causes the potential. Although we have in practice used it as tridimensional, it is expedient for the sake of the coming discussion to introduce its one dimensional form (2) restricted to the z direction (perpendicular to the galactic plane).

$$\partial^2\Phi/\partial z^2 = \partial K_z/\partial z = 4\pi G\rho \qquad (2)$$

The collisionless Boltzman equation governs the behavior in the phase space of any subsample under the influence of a predefined potential. Assuming axisymmetry and independence of radial and z motions, this equation can be written in its one dimensional form:

$$v_z \cdot \partial f_i/\partial z + \partial\Phi/\partial z \cdot \partial f_i/\partial z = 0 \qquad (3)$$

$v_z z$ is the z projection of the velocity, f is the distribution function of stars in the two dimensional phase space (v_z, z). The index i refers to the subpopulation under consideration, whatever its definition and irrespective of whether stars belonging to this subpopulation can or cannot be identified in the sky. Since velocity data are seldom sufficient to get more than the second order moment of the velocity distribution, the Boltzman equation is most often used through its velocity moments in the Jeans form:

$$1/\rho_i \cdot \partial(\rho_i \, \sigma_{iw}^2)/\partial z = -\partial\Phi/\partial z \qquad (4)$$

where σ_{iw}^2 is the variance of the velocity distribution in the z direction. Assuming that the symbol i refers now to a well identified isothermal subpopulation, there is a simple solution of equation (4):

$$\rho_i(z) = \sigma_i(0) \exp-((\Phi(z) - \Phi(0))/\sigma_{iw}^2) \qquad (5)$$

Life would be simple if we knew the mass distribution and just had to derive the resulting potential through equation (1), and then study the behavior of a given subpopulation in the phase space according to (3) or (5). But the density law is just what we are about and the only thing we can get directly is the apparent magnitude distribution of not quite well defined tracer populations. So we are facing first an inverse problem from magnitude counts to densities - our common experience is that it is a seriously ill-conditioned one - and then a double differentiation of the resulting curve to get the dynamical density. Most investigations so far, following the path opened by Oort (1932), try to tackle the problem directly. They get $\rho_i(z)$ and velocity dispersions for subsamples generally assumed to be isothermal or simple combinations of several isothermal components, solve (5) for

$\Phi(z)$ and double differentiate it according to (2) to get the local volume density.

Of course the situation would not be this bad if we knew at least the mathematical shape of the density law, or conversely the mathematical shape of the potential. Then any tracer population would just be required to constrain a few free parameters. No such magic formula has been found so far, but we can nevertheless try to put together anything we know about stellar distributions and mass distributions in practice, compute the resulting potential under several reasonable hypotheses, then question the tracer populations for the choice of the more realistic hypothesis. This is what statisticians call the *Bayesian approach*: facing an ill-conditioned problem, first use any *a priori* knowledge you may find.

TABLE I

Isothermal components used by Hill et al. (1979) and Bahcall (1984) to build their galaxy models

M_v	σ_{iw} (km/s)	$\rho_i(0)/\rho(0)$ (Hill et al.)	$\rho_i(0)/\rho(0)$ (Bahcall)
Main Sequence			
< 2.5	4	0.038	0.021
2.5-3.2	8	0.019	0.015
3.2-4.2	11	0.033	0.031
4.2-5.1	21	0.034	0.035
5.1-5.7	20	0.023	0.025
5.7-6.8	17	0.036	0.037
>6.8	20	0.262	0.298
White Dwarfs	21	0.185	0.052
H + He	5	0.287	0.469
H2 + Dust	3	0.083	
Giants	20		0.016
Halo	100		0.001

2. SOLUTIONS BASED ON GALAXY MODELS

This has been done by Hill, Hilditch and Barnes in 1979. They used a decomposition of mass contributors in the solar neighborhood as shown in Table I (taken from Bahcall 1984). Each contributor was given a relative local volume density and a local velocity dispersion. Hill et al. also adopted three different sets of solutions of the coupled Boltzmann - Poisson system proposed by Camm (1950). One solution was for isothermal things, others not, however Camm's solutions were for one self-gravitating component and so they are not likely to give a realistic shape for a multicomponent system. Camm's solutions are expressed in terms of z density and z dependence of the velocity dispersion of any tracer, thus they can be fitted to the observed density curves and velocity moments. Using their sample of A- and F-type stars based on the Upgren (1962) spectral type survey, Hill et al. made a number of trials with different assumed local volume densities. They found that $\rho(0) = 0.14$ best fitted their data. The point I would like to raise here is that the above quoted decomposition just constrains the mathematical shape of the solution, but assuming that this shape is realistic, the derivation of the local density is left to the tracer population.

John Bahcall (1984) proceeded somewhat differently. He adopted a mass decomposition very similar to Hill et al., but

1) he assumes that each component is isothermal.

2) He adds a non z-isothermal term to account for the halo contribution. This contribution is assumed constant in the relevant z range.

3) He introduces various tentative sets of unseen mass isothermal components.

4) Then he solves numerically the joint Poisson-Boltzmann equation for the potential of the whole composite system and so gets as many potentials as he has tentatively assumed different "Unseen Mass Disks".

$$\partial^2\Phi/\partial z^2 = 4\pi G\{\Pi\rho_i(0)\exp-((\Phi(z)-\Phi(0))/\sigma_{iw}^2)+(\Sigma_j \text{ UMD}_j)+\text{Halo}\} \quad (6)$$

Some of the main sequence sections appearing in Table I are obviously not isothermal and so are giants. This is not a major restriction since, as Bahcall has reminded us, any distribution can be represented to an arbitrary precision by a decomposition into isothermal components. So the decomposition is efficient as long as the right mass fractions are associated to the right velocity dispersions irrespective of which kinds of stars are associated to either quantity. With respect to Hill, Bahcall gains one degree of freedom: the unseen mass is not bound to be proportional to the seen one. Furthermore the solution is safely self-consistent. So he gets a set of realistic potentials and leaves it to tracer populations to decide which is the right one.

Given a sample tracer population, assuming that it is isothermal, and getting its velocity dispersion from other sources, it is a straightforward exercise to compute the predicted density distribution in a given potential. On the other hand, given the absolute luminosity distribution of the tracer stars and the apparent magnitude distribution of a sample towards one of the galactic poles, any inverse method including the m/logπ table due to Kapteyn provides an estimated "observed density law". So did Hill et al. using the F type star

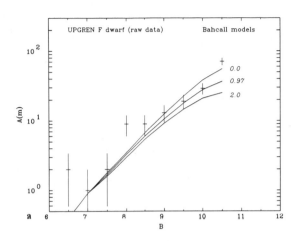

Fig. 1. Star count predictions based on Bahcall's density model under three different hypothesized unseen mass disks. Crosses with error bars are Upgren's data.

TABLE III.

Relative age distribution in some typical cells (absolute magnitude M_v, spectral types spt)

		AGE (10^9yr)											
MV	SPT	15	1.0	2.0	3.0	5.0	7.0	10.	12.	13.	14.	16.	
-2.0	3	1.0000	0.0000	0.0000	0.0000	0.0000	0.0000	0.0000	0.0000	0.0000	0.0000	0.0000	Supergiants
-1.5	13	0.0000	0.0000	0.0000	0.0000	0.0000	0.0000	0.0000	0.0000	0.0000	0.0000	1.0000	A.G.B EPII
-0.5	13	0.0000	0.0000	0.0000	0.0000	0.0000	0.0000	0.0000	0.3340	0.3330	0.3330	0.0000	A.G.B IPII
0.0	13	0.0000	0.0000	0.2169	0.2041	0.1315	0.2156	0.2102	0.0073	0.0072	0.0072	0.0000	Disc + IPII Giants
0.5	4	0.5958	0.4042	0.0000	0.0000	0.0000	0.0000	0.0000	0.0000	0.0000	0.0000	0.0000	Early Main Sequence
0.5	12	0.0000	0.1120	0.2672	0.1196	0.1314	0.0829	0.2540	0.0102	0.0101	0.0101	0.0023	Disc, IPII, EPII Giants
3.0	8	0.0215	0.9681	0.0101	0.0000	0.0000	0.0000	0.0000	0.0000	0.0000	0.0000	0.0000	MS
3.0	9	0.0000	0.0000	0.7757	0.2243	0.0000	0.0000	0.0000	0.0000	0.0000	0.0000	0.0000	MS
4.5	10	0.0048	0.1142	0.1590	0.1930	0.3997	0.1163	0.0000	0.0037	0.0037	0.0037	0.0017	Disc MS, IPII + EPII Subgiants
6.5	12	0.0000	0.0000	0.0000	0.0000	0.1184	0.3422	0.5287	0.0031	0.0031	0.0031	0.0014	Disc MS, IPII + EPII Subdwarfs
7.0	13	0.0000	0.0000	0.0000	0.0000	0.0945	0.2731	0.4219	0.0703	0.0701	0.0701	0.0000	Disc MS, IPII MS
7.5	13	0.0016	0.0679	0.0702	0.1005	0.1466	0.2351	0.3632	0.0046	0.0046	0.0046	0.0010	Disc + IPII + EPII MS
10.0	15	0.0016	0.0707	0.0731	0.1046	0.1587	0.2276	0.3638	0.0000	0.0000	0.0000	0.0000	Red Dwarfs

TABLE II.

Magnitude-spectral type distribution of disk stars in the solar neighborhood (stars/pc³)

MV	O7-O9	B0-B4	B5-B9	A0	A2-A3	A4-A9	F0-F4	F5-F9	G0-G7	G8-G9	K0-K4	K5-K6	K7-K9	M0-M4	>M5
-5.0	.46E-07														
-4.5	.60E-07	.60E-07													
-4.0		.20E-06													
-3.5		.40E-06													
-3.0		.62E-06													
-2.5		.10E-05													
-2.0		.20E-05													
-1.5		.35E-05													
-1.0		.66E-05	.60E-06												.18E-06
-0.5			.55E-05											.10E-05	
0.0			.13E-04	.28E-05	.37E-05							.14E-04		.39E-05	
0.5			.23E-04	.37E-05	.18E-04					.15E-04	.15E-04				
1.0				.50E-04	.90E-04	.14E-04				.16E-04					
1.5					.60E-04	.37E-04	.69E-05		.16E-04						
2.0						.60E-04	.28E-04	.69E-05	.93E-05						
2.5						.11E-03	.22E-03	.28E-04							
3.0							.37E-03	.16E-04	.84E-04		.83E-04				
3.5							.40E-03	.47E-03		.10E-03					
4.0								.11E-02							
4.5								.67E-03	.67E-03						
5.0									.16E-02						
5.5								.67E-03	.68E-03	.10E-02					
6.0											.18E-02				
6.5											.17E-02	.15E-03			
7.0											.14E-02	.17E-02	.59E-03		
7.5												.14E-02	.21E-02		
8.0													.12E-02	.12E-02	
8.5														.29E-02	
9.0	.60E-06	.60E-06												.36E-02	
9.5	.32E-05	.32E-05												.44E-02	
10.0	.57E-05	.57E-05												.50E-02	
10.5	.15E-04	.15E-04	.15E-04											.64E-02	
11.0		.25E-04	.25E-04	.25E-04											
11.5		.35E-04	.35E-04	.35E-04	.35E-04										
12.0		.52E-04	.52E-04	.52E-04	.52E-04										
12.5		.52E-04	.71E-04	.52E-04	.71E-04	.71E-04									.76E-02
13.0				.71E-04	.90E-04	.90E-04	.90E-04								.81E-02
13.5				.90E-04	.17E-03	.17E-03	.17E-03								.90E-02
14.0				.17E-03	.24E-03	.24E-03	.24E-03	.24E-03							.90E-02
14.5							.19E-03	.19E-03	.19E-03	.19E-03	.19E-03				.90E-02
15.0							.19E-03	.19E-03	.19E-03	.19E-03	.19E-03				.90E-02
15.5								.19E-03	.19E-03	.19E-03	.19E-03				.90E-02
16.0								.19E-03	.19E-03	.19E-03	.19E-03				.90E-02
16.5								.19E-03	.19E-03	.19E-03	.19E-03				.90E-02
17.0							.19E-03	.19E-03	.19E-03	.19E-03	.19E-03				.90E-02

samples selected by Arthur Upgren. They produced density laws which were later rescaled by J. Bahcall to more up-to-date absolute magnitudes. The comparison of the "observed" density to all model calculated ones provides the criterion to decide which model fits the data. That is what Bahcall (1984b) did and how he got the conclusion that the best fit model needed large amounts of unseen mass in the form of a disk. The density-density fit can be found in Fig. 1 of the quoted paper.

Note that Hill used Upgren counts down to m_B = 13 to derive the density as high as 500 pc while Bahcall limited his density-density fit to 200 pc because beyond this limit there is no more guarantee that Upgren's samples contain only members of the isothermal population. The contribution of stars farther than 200 pc is significant in all magnitudes beyond m_B = 9.5. Note also that, at this stage, it's a pity not to use the fact that we do know the computed density of each model to avoid the burden of the m/logπ table. It seems far simpler to use both density and absolute luminosity distributions to derive the computed apparent magnitude distribution resulting from each model and then search for the model that gives maximum likelihood to the magnitude count data.

Fig. 1 shows magnitude count predictions derived from some of the Bahcall's density models for a subpopulation assumed to represent Upgren's F5 - F9 stars. Predictions are limited to the useful range of magnitudes where stars farther than 200 pc do not contribute. We have plotted in the same figure the data from Hill et al. in the same range. Clearly enough there is little information in this range to help choosing between models without unseen mass (P = 0.0) and models involving as much unseen mass as seen (P = 1). Detailed analysis of this question can be found in Crézé et al. (1989).

3. SOLUTION BASED ON A GALAXY MODEL INVOLVING POPULATION SYNTHESIS

Now let me come to what we actually did in this context. Details of the following discussion can be found in Bienaymé et al. (1987).

3.1 Age Distributions in the HR Diagram

The basis of the approach is a model of population synthesis built by Annie Robin and myself (Robin and Crézé 1986). The model involves a scenario of star formation, evolutionary tracks and a few additional ingredients which make it possible to predict the distribution of solar neighborhood stars in terms of luminosity (resp. surface gravity), spectral type (resp. temperature), and age. In our model this distribution is materialized by a three dimensional table of which I can just show a few samples. Table II gives the absolute number density in the solar neighborhood for disk stars as predicted by this model, and Table III gives for a few cells taken in Table II the relative age distribution of stars found in the cell.

The distribution in Table II has been forced to fit Wielen's (1977) luminosity function marginally. The important point is that we can now produce a new decomposition of the visible mass in the solar neighborhood, not according to the absolute magnitude as Hill or Bahcall did, but according to the age. Then, gathering from the whole set of distributions the frequencies of stars with same age but different luminosities and colors (chemical compositions are also taken into account in the model), we get coeval subsets which are not homogeneous in spectral type, nor are they iso-absolute magnitude subsets. We cannot identify complete samples of this kind in the sky through any simple observable criterion, but we are sure that they do exist, and that stars from such subsets have experienced a common dynamical history. There is still to understand how this common but unknown dynamical history is reflected in the present kinematics of coeval samples.

3.2 Ages, Velocity Dispersions and Isothermal Components

The earliest explicit attempt we know of, to separate kinematic samples according to an age criterion, is due to Von Hoerner (1960). Von Hoerner identified areas in the HR diagram which admittedly would be coeval areas, at least the average stellar age was significantly different from one area to another. Von Hoerner concluded that the velocity dispersion substantially increases with increasing age. The age inhomogeneity of spectral type selected samples is also well established; it is clearly reflected in velocity distributions. Delhaye (1965) (see Fig. 2 of his paper), presents distributions for gK and dM samples with violent departures from normality; both categories are known to be age mixtures. A-type stars which are not so widely spread in ages show at least more regular distributions, but such samples may also include very young stars which are not in a steady state. The best age criteria have been used by Mayor (1974), Oblak (1983) and Carlberg (1985), based on accurate Strömgren photometry of F type stars and isochrones in the $M_v/\text{Log } T_{eff}$ plane. Wielen (1973) and Upgren (1978) used an independent calibration of ages based on the strength of Ca II emission lines (the HK Criterion from Wilson and Woolley 1970). It is remarkable that all these studies, however different and independent may be the age criterion, do produce compatible age versus velocity dispersion relations. There is yet no more evidence for significant residual inhomogeneity in the velocity distribution of coeval samples. Carlberg (1985) does provide a Kolmogorov test that no significant departure from normality can be found in his coeval samples. It has been proved several times (Mihalas 1968 Eqn 12.73, Bahcall 1984a) that in a stationary axisymmetric potential, there is a dynamical equivalence between the velocity distributions being Gaussian and the corresponding populations being isothermal (in the sense that $\partial(\sigma_{iw}^2)/\partial z = 0$). This is true at least within the limits of validity of the plane parallel hypothesis.

That is why we choose to rely upon the above described long accumulated knowledge of very local kinematic data, rather than trying to study the kinematics of a specific sample, for which spectral selection cannot be warranted at large distances from the Sun and for which an assumed isothermality can hardly be assessed due to limited statistical significance. Hence we use the age decomposition of all disk stars in the solar neighborhood to assign them velocity dispersions according to a well established age/velocity dispersion relationship. In practice we have adopted Mayor's (1974) relation, had we used Wielen's (1973) or Carlberg's (1985) things would not be very different.

3.3 Dynamical Consistency

Data from Tables II and III give the age frequencies for any spectral type and absolute magnitude in the solar neighborhood. Ages are linked to velocity dispersions through Mayor's (1974) relation. Hence predictions can be made of how many solar neighborhood stars of any type identified in the HR diagram, are associated to each isothermal component. Then, given a potential, the trend of the density law in the vertical direction can be obtained through equation (5). We search, amongst a family of realistic three-dimensional density laws, which one best mimics the above computed trend. Then the local mass density of each component can be associated to a density law which fits the trend of this component. Components can be added and completed by a few imposed contributors; the Halo and Dark Corona which obviously depart from the plane-parallel hypothesis and the interstellar matter which does not obey the collisionless dynamics. Unseen mass disks are eventually added at this stage. Finally, the sum of dynamically constrained and not-constrained mass contributors gives the total mass distribution that produces the potential according to equation (1). Details of the relevant 3D integration process are given in Bienaymé et al. (1987), appendices A and B.

Two iteration loops cooperate to achieve dynamical consistency. The vertical component of the new potential is used to recompute the density trends of all above quoted isothermal components, until the adopted set of density laws stabilizes. On the other hand, once a set is obtained, the resulting rotation curve is submitted to the constraints adopted by Caldwell and Ostriker (1981) and the Corona and Bulge densities are modified until these constraints are properly matched. Further iterations along the first loop are necessary since changing the Corona parameters does change the local vertical mass distribution. The whole process converges quickly and the third global run seldom brings significant changes. Thanks to this double constraint, scenarios involving unrealistic column densities can be ruled out.

Several self-consistent solutions of the 3-dimensional Poisson equation and 1-dimensional Jeans equation have been performed under different scenarios. One scenario assumes that the Galaxy mass is made of the visible contributors plus a low local density dark corona imposed by the rotation curve; others tentatively introduce unseen mass components. Each scenario produces a potential which is supposed to control the behavior of real stars in the phase space. There comes the difficulty of identifying tracer stars for which behavior can be both computed in potentials and observed in reality.

3.4   Constraints From General Star Counts

Most investigators so far have used photometric and/or spectroscopic criteria to identify tracer stars belonging to isothermal samples. The trouble is that, having properly identified in the solar neighborhood an isothermal family for which velocity dispersion has been accurately measured and for which absolute magnitude is safely calibrated, observations capable of producing large samples at faint magnitudes do not allow one to recognize family members. The important feature of the Bienaymé et al. (1987) approach is that, unlike Hill et al. (1979) or Bahcall (1984), isothermal components are not just artifacts linking mass contributors to scale heights; the population synthesis does predict how many stars from each part of the HR diagram are associated to each isothermal component. Hence under each assumed potential, we are able to compute the behavior of all stellar families currently encountered in the solar neighborhood. Thus we do not need to identify which observed stars belong to which isothermal family. We just have to check that the predicted distribution of magnitudes of all stars does fit the observed one. In Fig. 2, we show the fits of two self-consistent models compared to general star counts obtained by nine different authors, in different fields at high and intermediate galactic latitudes. The range of models compatible with the data has been estimated using a maximum likelihood scheme. The main conclusion is that no model involving large amounts of unseen mass can produce an acceptable fit. Unseen mass disks such as those introduced by Bahcall (1986) generate global shifts of the predicted magnitude counts which become discrepant with the bulk of the observed data. The fit is based on thousands of stars ranging from magnitude 6 to 22.

Depending on which range in apparent magnitude and which galactic direction we consider, very different stars are involved: some ranges are dominated by disk red dwarfs at moderate distances above the galactic plane, others by more distant F - G stars and giants. In magnitude bins beyond V = 16 Halo and/or Thick Disk stars make substantial contributions.

Different ingredients of the model may be responsible for such or such local discrepancy. Errors in absolute magnitude calibrations would eventually affect giants or part of the main sequence. Changes in the star formation scheme would produce more (or less) old stars in some part of the HR diagram, meaning more (or less) stars at large distances. Errors in photometric calibrations may result in a shift of all the data from one author with respect to others. A general conspiracy of all error sources to hide consistently the unseen mass is extremely unlikely.

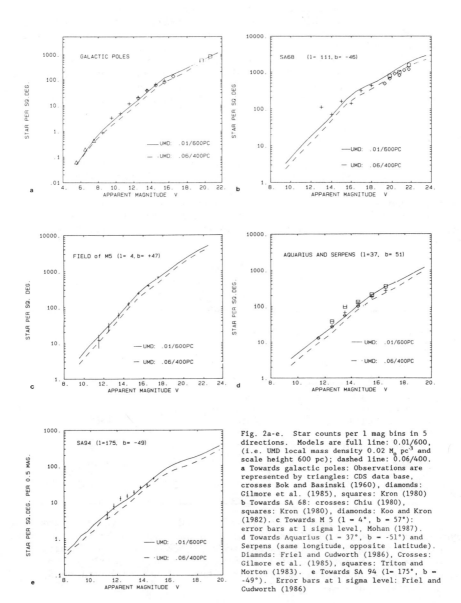

Fig. 2a-e. Star counts per 1 mag bins in 5 directions. Models are full line: 0.01/600, (i.e. UMD local mass density 0.02 M$_o$ pc^{-3} and scale height 600 pc); dashed line: 0.06/400. a Towards galactic poles: Observations are represented by triangles: CDS data base, crosses Bok and Basinski (1960), diamonds: Gilmore et al. (1985), squares: Kron (1980) b Towards SA 68: crosses: Chiu (1980), squares: Kron (1980), diamonds: Koo and Kron (1982). c Towards M 5 (l = 4°, b = 57°): error bars at 1 sigma level, Mohan (1987). d Towards Aquarius (l = 37°, b = -51°) and Serpens (same longitude, opposite latitude). Diamnds: Friel and Cudworth (1986), Crosses: Gilmore et al. (1985), squares: Triton and Morton (1983). e Towards SA 94 (l= 175°, b = -49°). Error bars at 1 sigma level: Friel and Cudworth (1986)

M. Crézé, A. C. Robin and O. Bienaymé

The apparent additional complexity does not come from additional hypotheses. Had we simply managed to observe as many independent isothermal samples as there are in the model, we would not bring a more stringent constraint. Of course, we do not compare the predicted behavior of each isothermal family to identified family members, but this is precisely what nobody can do. The bulk of stars involved lies between 50 and 1000 parsecs off the galactic plane, so that the trend of the gravitational force law is tightly constrained all over this range.

4. THE LOCAL MASS DENSITY

Table IV summarizes the local mass density characteristics of the self-consistent galaxy model adopted by Bienaymé et al. (1987). Error bars represent the limits placed on these characteristics by fitting general star counts. The surface density is actually a column density integrated to 1 kpc, which is the range where the data constrain efficiently the model and also the range where basic hypotheses remain valid.

TABLE IV

Local Mass Density Characteristics

	Volume Density $M_\odot$ pc^{-3}	Surface Density $M_\odot$ pc^{-2}
Interstellar Medium	0.040	11 $\pm$ 2
Total stellar disk	0.044	41 $\pm$ 4
Intermediate	negligible	0.16
Halo (Spheroid)	negligible	0.16
Unseen Mass Disk	0.01	5 $\pm$ 5
Corona	0.008	8 $\pm$ 2
Total Mass	0.102 $\pm$ 0.01	65 $\pm$ 10

There is obviously no room in this analysis for one more mysterious component besides the dark corona imposed by rotation curve considerations. So, using global modeling constrained by general star counts we get an estimate of the local mass density nearly one order of magnitude more accurate than previous investigations based on spectral type selected samples.

5. CONCLUSIONS

The main trouble with using the Boltzmann equation to estimate the local mass density and the force law perpendicular to the galactic plane comes from the difficult identification of isothermal samples out of the plane. There are two ways out of this difficulty. One has been followed by Bienaymé et al. (1987); it avoids sampling difficulties by gathering in a self-consistent galaxy model any a priori knowledge about stellar populations relevant to the problem, leaving only the local mass density to be tested against general star counts. The result is given in Table II: 0.102 ± 0.01 M_0 pc^{-3} for the local volume density and 65 ± 10 M_0 pc^{-2} for the surface density .

Another way towards the answer would be to get solutions of equation (3) for the distribution f_i of a tracer population in phase space under each assumed potential, then to observe both magnitudes and velocities of a tracer star sample and compare the computed f_i's with the observed density-velocity distribution. Provided the solution does not rely upon the isothermality hypothesis, this releases drastically the sampling difficulties, but imposes very large amounts of observing time since radial velocities should be observed for all stars in the sample. This has been done recently by Kuijken and Gilmore (1989) and will be developed by Gilmore later in this colloquium. The result is 0.1 M_0 pc^{-3} for the local volume density and 48 ± 8 M_0 pc^{-2} for the surface density. Both results rule out the dense unseen mass disks suggested by Bahcall's results (Bahcall 1984b). The long-lived discussion opened by Oort (1932) about the local missing mass is now restricted to a very narrow range. The star formation models developed by Larson (1986) to account for Bahcall's unseen mass become irrelevant (Robin et al. 1988).

Once we get rid of this local aspect of the missing mass problem, there are important questions still open. Essentially, star counts are the privileged link between local observations and the global galactic structure. This link would get much stronger if star counts were not only magnitude-color counts but involved also proper motions, giving some insight into the phase space distributions. Some new attempts in this direction will be described later in this colloquium, by Andreas Spaenhauer, based on the Basel program, and by Olivier Bienaymé and Annie Robin, based on the Besançon program. There are also limitations to connecting local things to global ones, arising from the not fully three dimensional modeling imposed by the plane parallel hypothesis. I suspect that Thomas Statler's contribution involving Stäckel potentials will open a promising window.

REFERENCES

Bahcall, J. 1984a Astrophys. J. 276, 156.
Bahcall, J. 1984b Astrophys. J. 276, 169.
Bienaymé, O., Robin, A. C. and Crézé, M. 1987 Astron.

Astrophys. 180, 94.
Camm, G. L. 1950 Monthly Not. Roy. Astron. Soc. 110, 305.
Caldwell, J. A. R. and Ostriker, J. P. 1981 Astrophys. J. 251, 61.
Carlberg, R. G., Dawson, P. C., Hsu, T. and VandenBerg, D. A. 1985
    Astrophys. J. 294, 674.
Chiu, L. T. G. 1980 Astrophys. J. Suppl. 44, 31.
Crézé, M., Bienaymé, O. and Robin, A. C. 1989   Astron.
    Astrophys. 211, 1.
Delhaye, J. 1965 in Galactic Structure, Stars and Stellar Systems
    Vol 5, A. Blaauw and M. Schmidt, eds., University of Chicago Press,
    Chicago, p. 61.
Friel, E. D. and Cudworth, K. M. 1986 Astron. J. 91, 293.
Gilmore, G. 1984 Monthly Not. Roy. Astron. Soc. 207, 223.
Gilmore, G., Reid, N. and Hewitt, P. 1985 Monthly Not. Roy. Astron.
    Soc. 213, 257.
Hill, G. Hilditch, R. and Barnes, J. V. 1979 Monthly Not. Roy.
    Astron. Soc. 186, 813.
Koo, D. C. and Kron, R. 1982 Astron. Astrophys. 105, 107.
Kron, R. 1980 Astrophys. J. Suppl. 43, 305.
Kuijken, K. and Gilmore, G. 1989 Monthly Not. Roy. Astron. Soc.,
    in press.
Larson, R. B. 1986 Monthly Not. Roy. Astron. Soc. 218, 409.
Mahon, V. 1987, private communication.
Mayor, M. 1974 Astron. Astrophys. 32, 321.
Oblak, E. 1983 Astron. Astrophys. 123, 328.
Oort , J. 1932 Bull. Astron. Inst. Netherlands 6, 249.
Robin, A. and Crézé, M. 1986 Astron. Astrophys. 157, 71.
Robin, A. C., Crézé, M. and Bienaymé, O. 1988 in XXIIIth
    Rencontre de Moriond, Dark Matter, J. Audouze and J. Tran Thanh
    Van, eds., p. 239.
Tritton, K. P. and Morton, D. C. 1984 Monthly Not. Roy. Astron. Soc.
    209, 429.
Upgren, A. R. 1962 Astron. J. 67, 37.
Upgren, A .R. 1978 Astron. J. 83, 626.
Von Hoerner, S. 1960 Mitt. Astron. Rechen Inst. Heidelberg, Ser. A ,
    13.
Wielen, R. 1974 in Highlights of Astronomy, Vol 3, G. Contopoulos,
    ed., Reidel, Dordrecht, p. 395.
Wielen, R. 1977 Astron. Astrophys. 60, 263.
Wilson, O. J. and Woolley, R. 1970 Monthly Not. Roy. Astron. Soc.
    148, 463.

DISCUSSION

SCHECTER: Is the range of absolute magnitude at a given spectral type an observed spread or a theoretical spread due to a range of metallicities?

CREZE: It is essentially an effect of the spectral type resolution of the evolutionary model.   The model produces stars of different masses at different epochs, then stars follow different evolutionary tracks. Stars of different masses and ages happen to fall in the same spectral type bin but are not at the same evolutionary stage and have different surface gravities.   Apart from halo stars, which are dealt with separately, metallicity effects are second order ones at this level.

RATNATUNGA: How were the theoretical luminosity functions normalized to fit Wielen's observed luminosity function?

CREZE: The whole distribution in the HR diagram has been summed over spectral types and gives an approximation of $F(M_v)$ and then frequencies in each spectral type bin were scaled using the factor $F(M_v)$Wielen/$F(M_v)$.   The relative age frequencies have been kept unchanged.

CHIU: In your Table III, relative age, there are two odd numbers. What do they represent?

CREZE: This just means that all stars in the relevant bin of the HR diagram have the same age.

KING: In this problem, the density distribution and the velocity distribution are equally important.   Could you make clearer your data, or assumptions, about velocities?

CREZE: Once you have obtained age distributions in all parts of the HR diagram and once you admit that coeval populations are isothermal you just need the age/velocity dispersion relation.   This was taken from Mayor (1974).   Since the star counts we use do not contain velocity information, there is no need for more.   On the other hand we have checked that results were robust to realistic errors in the Mayor's relation.

MARAN: Is it conceivable that the local distribution which appears to require dark matter can have been "stirred up" by an ancient tidal interaction, for example with a passing globular cluster?

CREZE: The tracer samples we use are dominated by stars with relatively high velocity dispersions - such populations are stable against local perturbations.

RATNATUNGA: What statistical significance can you place on the fraction of missing disk-like mass using your fits to star counts as a function of apparent magnitude?

CREZE: Trying to give 1 sigma limits I would say 0.01 ± 0.01 M_o pc^{-3} for the volume density or 4 ± 4 M_o pc^{-2} for column density.   Since many error sources are not exactly random errors, in the 1987 paper we preferred to give upper and lower limits.   There we get something between 0 and 0.03 in volume density.

STATISTICS OF GALACTIC DISK TRACER STARS

Andrew Gould

Institute for Advanced Study

ABSTRACT: It is shown that low-altitude ($\lesssim$ 1 kpc) tracers are far better suited to the determination of the missing mass in the galactic disk than are high-altitude tracers. A specific strategy is given for optimizing tracer observations and analysis. The observations should span roughly three scale heights.   Uncertainties from the velocity measurements dominate all other uncertainties and this implies that velocities should be measured on all stars observed, that the integration time should be long enough that the error is instrument-limited, and that repeat observations should be done at least one year later.   Typically, the stars should be grouped in four altitude bins.   A few hundred stars are sufficient to distinguish with 3-σ confidence between no disk missing mass and half missing mass.   To distinguish between hot and cold missing mass requires $\sim$ 7000 stars from one survey or $\sim$ 1000 stars from each of two surveys of tracers with substantially different velocity dispersions (e.g. F dwarfs and K dwarfs).   These figures include corrections of a factor of $\sim$ 3 due to the fact that the tracers may deviate from being isothermal and still be measured as isothermal within errors.

1.   INTRODUCTION

I will discuss the statistics of tracers of the galactic disk potential with a view toward defining an optimal observation strategy for low-altitude tracers. By "low altitude", I mean $\leqslant$ 1 kpc. My talk will be divided into two sections. First I will show why it is that low- (as opposed to high-) altitude tracer observations are the indicated method for measuring the missing mass in the disk.   Then I will give a series of prescriptions for optimizing such observations and give general estimates for how many tracers must be observed in order to determine the magnitude and velocity dispersion of the missing mass.   I will not attempt to give detailed derivations of any of my conclusions.   Those interested in such details may read about them in two preprints (Gould 1989a, 1989b) which can be obtained from me after the talk.

A. G. Davis Philip and P. K. Lu (eds.)
The Gravitational Force Perpendicular to
the Galactic Plane   19 - 28
©   1989   L. Davis Press

2. WHY LOW-ALTITUDE TRACERS

It seems obvious to many that the best way to determine the missing mass in the disk is to observe tracers high above the plane. From such tracers one may infer the column density of the disk. Then, knowing how much mass is truly in the disk and how much would be in it if only the observed components were actually present, one could presumably tell how much mass was missing. To help understand why this chain of reasoning breaks down, I will compare low- and high-altitude tracers in four respects. First, in terms of parameterization of the potential. Second, in terms of how uncertainties in the large-scale structure of the galaxy affect the data analysis. Third, as to which is better at measuring the column density, and finally, as to which is better at measuring the missing mass.

2.1 Parameterizing The Potential

The potential at high altitudes can be parameterized quite simply in terms of a linear term K and a quadratic term F,

$$\psi(z) = -\psi_0 + Kz + Fz^2. \qquad\qquad 2.1$$

Here K is proportional to the column density of the disk, $K = 2\pi G\Sigma_0$, and represents the gravitational force of the disk (above the bulk of the disk matter distribution). I call F the "large-scale structure constant" because it is related to the solar galactocentric distance, the solar rotation velocity, the radial scale of the disk and the ellipticity of the halo. It is sometimes expressed in terms of an "effective halo density", $F = 2\pi G\rho_{\text{eff}}$. The constant ψ_0 has no physical meaning and is incorporated only to meet the usual normalization and symmetry conditions

$$\psi(0) = \psi'(0) = 0. \qquad\qquad 2.2$$

The simple form (2.1) allows one to represent all possible high-altitude potentials with only two parameters. This is one of the features which make high-altitude tracers appear so attractive.

By contrast, the low-altitude potential is a very complicated, non-linear function of its parameters. Following Bahcall (1984) we can specify the velocity dispersions $\overline{v_i^2}$ and densities at the plane $\rho_i(0)$ of various components of the disk, including both those observed and also possible additional unobserved components. One may then solve self-consistently the Poisson-Boltzmann equation

$$\frac{1}{4\pi G}\frac{d^2\psi}{dz^2} = \sum_i^N \rho_i(0)\exp[-\psi(z)/\overline{v_i^2}] + \rho_{\text{eff}}. \qquad\qquad 2.3$$

Here ρ_{eff} is the same effective halo density mentioned above.

2.2 Uncertainties From The Large-Scale Structure

Exactly because the high-altitude potential is so simple, containing as it does only two parameters, its analysis is affected very seriously by uncertainties in the large-scale structure of the Galaxy. If one attempts to use tracer data to fit the potential to the functional form (2.1), one finds that the best fits for K and F are highly anti-correlated. In fact, in a typical model which I worked out in detail, I found a correlation coefficient of -97%. This means that it is almost impossible to obtain any significant constraint on K without making some assumption about F. Fortunately, one does have some independent information about the large-scale structure of the Galaxy, namely that Sun is held in its galactic orbit by a combination of the disk and halo masses. (I use the term "halo" to refer to all the non-disk components of the Galaxy - bulge, spheroid, and halo.) If one assumes that the disk has an exponential radial profile (a reasonable assumption justified by the light curves of other spirals) and that the halo is spherically symmetric, one may show that F and K obey a linear relation which is a function only of the Solar Galactocentric distance and Solar orbit speed. (There is also a very slight dependence on the radial scale length of the disk, but this is of no practical importance.) This relationship may be evaluated numerically using galactic parameters from Binney and Tremaine (1987, p. 13-14),

$$\frac{F}{F_\star} + 0.52\frac{K}{K_\star} = 1.52 \pm 0.35 + 2\epsilon, \qquad 2.4$$

where ϵ represents the ellipticity of the halo and is for the time being taken to be zero. The normalization $K_\star$ is the value of K assuming no missing mass and $F_\star$ is the expected value of F assuming that $K = K_\star$,

$$K_\star \sim 1.3 \; (\text{km s}^{-1})^2\text{pc}^{-1}; \qquad F_\star \sim 2.2 \times 10^{-4} \; (\text{km s}^{-1})^2\text{pc}^{-2}. \qquad 2.5$$

There are two important points to note about equation (2.4). First, the fractional error in F is rather large and second, K and F are again anti-correlated. The physical reason for this anti-correlation is the same as it was in equation (2.1): one does not know whether the measured gravitational force is due more to the halo and less to the disk or whether the reverse is true. One would like to say that equation (2.4) gives us F to within 35% and then plug this 35% uncertainty into equation (2.1). Unfortunately, the fact that both equations anti-correlate K and F means that such a procedure would substantially underestimate the errors. Rather, one must adopt a more formal approach of combining covariance matrices.

In addition, there is no reason to believe that the halo is exactly spherical, and indeed, no data of which I am aware which constrain significantly its ellipticity. Thus it is possible that ϵ is ~0.1 or 0.2 or some other value. For this reason, equation (2.4) puts much weaker constraints on equation (2.1) than would be the case if we knew the ellipticity. It is clear then that our lack of knowledge about the large-scale structure of the Galaxy seriously compromises our ability to interpret data from high-altitude tracers.

By contrast, the uncertainty in F has almost no effect on the interpretation of low-altitude data. Although the effective halo density does enter equation (2.3), there are many other *known* parameters (the densities and velocity dispersions of the observed components) which also enter and which dominate the potential, thereby diluting the effect of our ignorance of F. The scale of low-altitude measurements is just too small for the large-scale structure of the Galaxy to play much role.

2.3 Determining The Column Density

From what was said in the previous section, it is clear that high-altitude tracers would be a good way to determine the column density if the ellipticity of the halo could be constrained. Under these conditions one could determine K to within 25% with 350 tracers, 16% with 1000 tracers or 10% with 3500 tracers (all at the 1-σ level). (See Gould 1989a, section V.) These figures would improve substantially if better measurements could be made of the Solar Galactocentric distance and Solar speed.

By contrast, low-altitude tracers cannot by themselves be used to measure the column density for the simple reason that a dark high-velocity component would be virtually undetectable by low-altitude tracers. Such a dark component could nevertheless add substantially to the disk mass.

2.4 Determining The Missing Mass

There are several reasons why high-altitude tracers are an extremely poor means of measuring the missing mass in the disk. First, they directly measure only the column density, not the missing mass itself. To understand the difference, consider the following example. Suppose that a set of tracer measurements showed that K was restricted to the range (at 1 σ)

$$K = 1.2K_* \pm 25\%. \qquad\qquad 2.6$$

(The error is that associated with ~ 350 tracers. The central value is my estimate of what the Kuijken and Gilmore (1989) data would show if equation (2.4) were incorporated into their analysis.) One might think that one could infer from this equation (at the 1-σ level) that no more that one third of the matter in the disk was in dark components. However, because of non-linear interactions between the dark and observed components, this naive conclusion applies only to dark matter in a very hot component. For a dark component of "average" velocity dispersion (~ 18 km/s), equation (2.6) is consistent with up to *half* the disk being dark, and for a cold dark component, the figure is more like two thirds. Moreover, the actual errors which may be expected from any survey in the near future are substantially larger than those quoted above because of the uncertainty in the halo ellipticity. Thus, even by observing a very large number of high-altitude stars, it is difficult to put any significant constraints on missing disk matter.

The situation with respect to low-altitude tracers is much more favorable. If one ignores the possibility that a large fraction of the dark matter is in a very-high-velocity component ($\gtrsim$ 40 km/s) then one may distinguish between half the disk being in a dark component and no dark component with 3-σ confidence using less than 200 *well-chosen* stars. I emphasize "well-chosen" because a poor observation strategy can degrade the effectiveness of a low-altitude survey by a factor of 2 or even more.

3. STRATEGY FOR LOW-ALTITUDE OBSERVATIONS

One may learn a great deal about the optimal strategy for low-altitude tracer observations by making a detailed study of the statistics. While I cannot, in a talk of this type, give a detailed description of the statistics, I can summarize the basic conclusions.

The first and perhaps most important point is that one should strive to make $\lambda \sim 1$, where λ is the *effective scale height* of the set of tracer observations. Formally, λ is defined in terms of three quantities: $\psi(z)$, some canonical potential (for example, the no-missing-mass potential); A(z), the physical area observed as a function of altitude (for an area complete survey A(z) α z^2); and $\overline{v^2}$, the measured velocity dispersion of the tracers. One may then define a "normalized" potential, $\hat{\psi}(z) \equiv \psi(z)/\overline{v^2}$. The effective scale height is defined as the dispersion of the normalized potential over the volume observed,

$$\lambda^2 \equiv \left\langle \hat{\psi}^2 \right\rangle - \left\langle \hat{\psi} \right\rangle^2,\qquad 3.1$$

where

$$\left\langle \hat{\psi}^n \right\rangle \equiv \frac{\int dz\, A(z) e^{-\hat{\psi}(z)} [\hat{\psi}(z)]^n}{\int dz\, A(z) e^{-\hat{\psi}(z)}}.\qquad 3.2$$

A rough rule of thumb is that an area complete survey to three physical scale heights has $\lambda \sim 1$.

The effectiveness of the survey per star increases linearly with λ until $\lambda \sim 1$, but remains basically constant after that. That is why it is so important to achieve an effective scale height of *at least* one. One may push λ up above one by observing stars at still higher altitudes but this does not improve the effectiveness of the survey and tends to squander a lot of telescope time on faint stars. Thus pushing λ above one may be not only wasteful, but counterproductive. Beyond these considerations, if stars of sufficiently high altitude are observed, then a knowledge of F, the large-scale structure constant is required to interpret the data.

The second point is that the statistical errors are dominated by uncertainties from velocity (as opposed to distance) measurements. The dominance is overwhelming when $\lambda \gtrsim 1$. This fact has a number of important implications for tracer observations. First, one should obtain spectra on all stars in the sample. For a particular observational program, it might be easier to do the photometric (distance) measurements than the spectroscopic (velocity) measurements, and so one may be tempted to "improve" the overall effectiveness of the survey by doing a lot of photometry in a short period of time, thus driving down dramatically the distance part of the error. However, since the distance uncertainties play only a small role, such a course is actually counterproductive. Next, one should integrate on each star long enough so that the velocity measurements are instrument- (as opposed to photon-count-) limited. Again, it may seem like a waste of time to sit on an individual star so long, but the number of *repeat measurements* (see below) required is directly related to the instrumental precision. Finally, because velocity errors play such a major role, it is very important to plan repeat measurements carefully. It is generally understood that the measured velocity dispersion has two contributions, one from the true dispersion of velocities of the stars observed and the other from the dispersion in the instrument. While one may estimate the latter from the instrument specifications, the velocity errors play such a large role in the analysis that it is necessary to actually measure the instrument dispersion by doing repeat measurements. One may show that the optimal number of such measurements is

$$\frac{N_{\text{repeat}}}{N_{\text{stars}}} = \left(\frac{\overline{u^4} - \overline{u^2}^2}{\overline{v^4} - \overline{v^2}^2} \right)^{1/2} \sim \frac{\overline{u^2}}{\overline{v^2}}, \qquad 3.3$$

where $\overline{u^2}$ is the instrument dispersion and $\overline{v^2}$ is the total measured dispersion, and where the final estimate was made on the basis of Gaussian statistics. In actual fact, there is not one source of spurious velocity dispersion, but two. Some of the dispersion will be due to motion about unseen companions. There is no way to estimate this effect accurately on the basis of current data and it would be impossible to find all the binaries which contribute to it. However, one may measure the total effect by doing the repeat measurements one year or more apart. If one wished to determine the separate contributions from binaries and from the instruments, it would be necessary to do two sets of repeat measurements, one the next night and one a year later. However, since one is interested only in the combined effect and since one does not know a priori which effect is larger, one should do all repeat measurements a year later. Next-night repeat measurements are virtually a complete waste of time.

The third point is that there are two principal independent numbers which may be extracted from the velocity data and both must be included explicitly in the analysis. One quantity, as is well known, is the velocity dispersion. The other seems less well known, but is *twice* as important. I call it τ^2, the thermal dispersion. While I cannot give the technical definition of the thermal dispersion here, the basic idea is as follows. Suppose that the velocities are actually drawn from *two* Gaussians which are constrained to have equal numbers of stars each. Let the velocity dispersions of these two components be $\overline{v^2}_{high}$ and $\overline{v^2}_{low}$. Then

$$\overline{\tau^2} \sim \left(\frac{\overline{v^2}_{high} - \overline{v^2}_{low}}{\overline{v^2}_{high} + \overline{v^2}_{low}} \right)^2 \qquad\qquad 3.4$$

(In general, any distribution can be decomposed into a sum of several Gaussians, although the coefficients need not be positive as they were in the above example. The general definition of τ^2 takes account of this.) Since the thermal dispersion is essentially independent of the velocity dispersion, it can be represented in the analysis by its own random variable. And since uncertainties in the thermal dispersion are twice as important as uncertainties in the velocity dispersion (see Gould 1989b, Figs. 1 and 2), it must be treated as an additional independent source of error. That is, it cannot be treated as a perturbation on the other "main" sources of error. (I should point out that τ^2 can be measured by two independent statistics. One is the kurtosis of the distribution and the other is the correlation of the velocity dispersion with the potential. Mathematically, it is possible for τ^2 to be negative, although this is not apparent from the simplified definition (3.4) given above. Physically, I know of no reason why it should not be negative, so this possibility must be permitted. If, for example, $\tau^2 = 0.0 \pm 0.1$, the uncertainty in the thermal dispersion in the negative direction cannot be ignored.)

The fourth point is that the data should be binned in ~ 4 altitude bins. The density distribution gives a nice "picture" of the (logarithm) of the potential and so one is tempted to plot lots of points on this picture in order to have a smooth curve. However, a detailed analysis shows this is not the optimal way to represent the data. The physical reason for this is that each additional bin contributes to the uncertainty, but only gives additional information if there is enough data to resolve structure in the potential on the scale of the bin number. A few hundred, or even a thousand stars is not enough to resolve structures smaller than 4 bins.

Finally, I should like to conclude by giving estimates of the number of stars required to reach various types of conclusions (at the 3-σ level). Suppose that half the mass of the disk were in unobserved components. Then less than 200 stars (chosen according to the above criteria) would be required to rule out the possibility that there was no missing mass. Suppose that this missing component was fairly hot, say 25 km/s. Then 7000 stars would be required to rule out the possibility that the dark component was extremely cold, that is ~ 4 km/s. Here I mean 7000 stars from a single survey. Such a large number of observations would appear out of reach for the present. However, if two surveys were combined, one of relatively cold tracers at ~ 10 km/s, and the other of relatively hot tracers at ~ 20 km/s, then this number could be reduced substantially.   In this case, the same goal could be achieved with only 1000 stars from each survey.   And reporting on this synergy seems an appropriate way to end an address to a conference which brings together diverse workers on the K_z problem.

REFERENCES

Bahcall, J. N.   1984   Astrophys. J. **276**, 156.
Binney, J. and Tremaine, S.   1987   Galactic Dynamics, Princeton
    University Press, Princeton.
Gould, A.   1989a   Astrophys. J., in press.
Gould, A.   1989b   Astrophys. J., to be submitted.
Kuijken, K. and Gilmore, G.   1989   Monthly Not. Roy. Astron. Soc.,
    in press.

DISCUSSION

KING: You have given a good discussion of statistical problems, but I want to emphasize that a major concern, throughout the history of this problem, has been systematic sources of error. With regard to velocities: (a) measurement errors of good modern radial velocities are negligible in this problem; (b) you should also have proper motions, which would add tremendously and offer some checks on systematic errors.

GOULD: It is not necessary for the velocity measurements to reach state-of-the-art limits, only that the "errors on the errors" be small compared to the statistical uncertainties. I have ignored proper motion studies because I believed that the errors on errors were not known (see above sentence). I am learning here that this concern may be outdated.

SCHECHTER: Ivan, your demand for km/s accuracy is hard to achieve for non F dwarf non K giants at a kiloparsec. It would be hard to do at Lick.

STATLER: To amplify Ivan King's remark about systematic effects, the low-z tracers are going to be particularly susceptible to systematic errors stemming from the non-axisymmetry and the non-up-down symmetry of the disk. There are things like the missalignment of the molecular and stellar disks, the warp, spiral arms, and the local bubble in the ISM, each of which can perturb the velocity dispersion at a ~10% level.

GOULD: I agree, and I'm glad you have been able to estimate these effects quantitatively.

GILMORE: I am a bit puzzled that you suggest so few distance bins. Would not a Nyquist sampling rate of ≥ 2 bins/distance over which one expects structure in the potential be better? Most of the demand for very large numbers of stars in your analysis comes from the problem of distinguishing realistic "isothermal" deconvolutions of binned data. Why not advocate an analysis of the observed distribution function? By avoiding moments one in effect uses the data more efficiently. And lastly have you calculated an error analysis using your method for the Kuijken/Gilmore data?

GOULD: Actually this type of Nyquist analysis gives roughly the right answer. As I showed, one can hope to extract one or perhaps two pieces of information, the missing mass and its velocity dispersion. One or two degrees of freedom means 2 or 3 bins. As it happens, a detailed analysis gives 4. Secondly, actually, the effect is the reverse of what you imply. Suppose the stars really were drawn from 2 Gaussians. Then my analysis would give the exact errors. If we have less information than that, the errors are worse, though I believe not much worse. The Kuijken/Gilmore approach cannot be used for low-altitude tracers because the extreme high-altitude behavior is unknown. I thought you pointed this out in your MNRAS paper. Finally, if one assumes that the distance error is negligible compared to the velocity error, then the uncertainties in K and F (without the Galactic-rotation constraint) are roughly 30% and 45% respectively. The latter value is rather troubling because the center of your K - F ellipsoids is low in F by 100%. Ignoring this and simply combining covariance matrices, I think the error in K will be reduced substantially, perhaps to 20%. However, I also think the central value of K will rise perhaps to 1.2 - 1.3 of the no-missing-mass value because the statistical error in F is now smaller than the error in the rotation constraint.

FLYNN: You have argued for low altitude tracers - but the major error here is uncertainty in the thermal dispersion. This error would be reduced by high altitude tracers - right? Or does the maximum information about this come at the turnover point at $\lambda = 1$ (effective scale height)?

GOULD: There are two independent sources of information about thermal dispersion: the kurtosis and the potential-velocity dispersion correlation. The latter does improve with increasing λ but the improvement saturates above $\lambda \sim 1$.

PHOTOMETRIC AND KINEMATIC STUDY OF F- AND G-STARS AT THE SGP: PRELIMINARY RESULTS

Ch. F. Trefzger

Astronomical Institute, University of Basel

J. W. Pel

Kapteyn Laboratory

M. Mayor and M. Grenon

Geneva Observatory

A. Blaauw

Kapteyn Laboratory

1. INTRODUCTION

This study aims at determining the stellar metallicity distribution and radial velocity dispersion for F-G stars in the local disk-halo transition region. In the high-latitude fields SA 141 (South Galactic Pole), SA 94 and SA 107 a sample of 251 F - G stars was selected using the photographic RGU photometry of the Basel halo survey. The photometric selection criterion of $(G - R) < 1.15$ (approximately $(B - V) < 0.7$) corresponds to a lower temperature limit of 5600 K. Since the blanketing lines in the RGU two-color diagram are almost vertical, this temperature selection is nearly independent of stellar metallicity. The completeness limits of our sample are as follows: SA 94: $V = 13.9$ (93 stars), SA 107: $V = 14.3$ (46 stars) and SA 141: $V = 14.8$ (112 stars).

2. PHOTOMETRIC DETERMINATION OF DISTANCES AND METALLICITIES

The observations were carried out with the Walraven five-channel photometer and the 91 cm Dutch telescope at ESO, La Silla during several runs starting in 1981. The effective wavelengths and bandwidths of the La Silla VBLUW passbands are (in Angstroms): V:5441 (708), B:4298 (423), L:3837 (221), U:3623 (232), W:3236 (157). The

A. G. Davis Philip and P. K. Lu (eds.)
The Gravitational Force Perpendicular to
the Galactic Plane   29 - 32
©   1989   L. Davis Press

VBLUW data for the program stars have been published by Pel et al. (1988); the extensive data on calibration stars will soon be published separately.

Assuming negligible or at most small foreground interstellar reddening in the directions of the high latitude fields investigated, VBLUW photometry enables us to separate the effects of temperature, metallicity and gravity for intermediate- type stars (Lub and Pel 1977, Trefzger, Pel and Blaauw 1984). The color index (V - B) is primarily a measure of T_{eff} and the colors (B - L) and (L - U) are very sensitive indices for [Fe/H] and log g, respectively. The calibration of these indices was made semiempirically using both theoretical colors based on the Kurucz model atmospheres (Kurucz 1979 and unpublished) and observed colors for many stars from the catalogue of [Fe/H] determinations by Cayrel et al. (1985). All calibrations were made differentially with respect to the Hyades main sequence. The physical parameters of our program stars are in the following ranges: $5350 < T_{eff} < 7200$, $-1.0 <$ [Fe/H] < 0.3 and $3.7 < \log g < 4.5$. This means that the majority of the stars are moderately metal deficient dwarfs with only a few subgiants in the sample. In order to determine absolute magnitudes and distances, we established an empirical temperature-absolute magnitude relation based on those stars in the catalogue of Cayrel et al. (1985) that have known trigonometric parallaxes and whose photospheric parameters are in the same range as those of the program stars.

The absolute magnitudes determined in this way vary between $M_v =$ $+3.2$ and $M_v = +6$. The completeness limits in the three observed fields correspond to distances from the galactic plane of 580 pc for SA 141, 300 pc for SA 107 and 290 pc for SA 94.

3. DISTRIBUTION OF STELLAR METALLICITIES AND RADIAL VELOCITIES

The distribution of stellar metallicities versus distance from the galactic plane for the combined sample shows a rather large range in [Fe/H] at any given distance, much larger than the observational accuracy. The mean [Fe/H] is around -0.1 at z = 100 - 200 pc and it drops to -0.5 for stars in the interval 600 pc < z < 1200 pc. The distribution cannot be represented satisfactorily by a general metallicity gradient or a two-component model with metal-rich disk stars and metal-poor halo stars. The observations are more adequately described in terms of a superposition of three stellar populations with different metallicity distributions, as proposed by Gilmore and Wyse (1985). A more detailed discussion of this point is in progress.

During two observing runs in 1986 and 1987 radial velocities were measured in SA 141 (the SGP field) for a total of 55 stars down to V = 13.5 using Coravel on the Danish 1.5 m telescope at La Silla. Even with this very limited kinematic data some trends can already be noted. The velocity dispersion increases from 17 km/s for stars with z < 300 pc to a value of 25 km/s for stars in the distance interval 300 pc < z

< 650 pc. This is comparable with the dispersions found by Gilmore and Wyse (1987) for K dwarfs in this distance range. The variation with stellar metallicity is more pronounced: for [Fe/H] near 0.0 the velocity dispersion is 11 km/s, whereas it increases to about 28 km/s for the stars with [Fe/H] around -0.5.

4. SUMMARY

A sample of 251 F-G stars at high galactic latitudes was selected from the Basel halo survey using a photometric selection criterion corresponding to T_{eff} > 5600 K. The stellar temperatures, metallicities and surface gravities are determined from Walraven VBLUW photometry. Most stars turn out to be dwarfs. The metallicity distribution up to z = 1200 pc supports the presence of a stellar component with intermediate metallicity. Radial velocities have been measured for about half of the program stars in SA 141.

ACKNOWLEDGEMENTS

We gratefully acknowledge financial support by the Swiss National Science Foundation and are very grateful to Dr. M. Grenon for spending a lot of time and effort on the Coravel observations.

REFERENCES

Cayrel de Strobel, G., Bentolila, C., Hauck, B. and Duquennoy, A.   1985
    Astron. Astrophys. Suppl. 145.
Gilmore,G. and Wyse, R. F. G. 1985 Astron. J. 90, 2015.
Gilmore, G. and Wyse, R. F. G. 1987 in The Galaxy, G.Gilmore and
    R. F. Carswell, eds., Reidel, Dordrecht, p. 247.
Kurucz, R. 1979 Astrophys. J. Suppl. 40, 1.
Lub, J. and Pel, J. W. 1977 Astron. Astrophys. 54, 137.
Pel, J. W., Trefzger, Ch. and Blaauw, A. 1988 Astron. Astrophys.
    Suppl. 75, 29
Trefzger, Ch., Pel, J. W. and Blaauw, A. 1984 The Messenger No.
    35, p. 32.

DISCUSSION

YOSS: Do you correct for reddening for your cut off limit of 1.15 as a function of galactic latitude?   What is the spectral range for your [Fe/H] frequency distribution.

TREFZGER: No, we did not correct our cut off limit, because reddening is so small that it does not show up in the photographic colors.   The spectral range is F - G5 or G6.

FLYNN: What is the mean or median abundance of "thick disk" stars in -0.8 < [Fe/H] < -0.4?

TREFZGER: Mean of [Fe/H] = -0.6.

RATNATUNGA: In your analysis you assumed an intermediate component with a scale height of 1300 pc and also derive a normalization of 11%. Since your samples are not very sensitive to scale height could the intermediate component that you see be an old disk of scale height 650 pc rather than the thick disk particularly since you find a velocity dispersion of 28 km/s for these stars?

TREFZGER: Since the interpretation of our data in terms of absolute magnitudes and hence distances is still preliminary, I would hesitate at the present time to draw conclusions of this kind; the only thing that we can say is that a component with intermediate metallicity is needed to describe the observations.

WING: What is the width of the L band at 3850 Å which is used for measuring metallicity? This will determine the relative importance of Fe lines, the calcium K-line, and the CN (0,0) band which is present at solar temperatures.   Statistically, these absorbers are strongly correlated, but individual stars can be found that depart from these correlations.

TREFZGER: The width is on the order of 200 Å. We assume that for dwarf stars we do not have CN-anomalies that affect our metallicity estimates.

NORRIS: I was interested in the stars in your (B-V) histogram for -0.8 < [Fe/H] < -0.4, and in particular the number of stars in the 0.4 - 0.5 bin, which are presumably younger than 47 Tuc. I have two questions (1) what is the error in your determination of (B-V), and (2) what happens to the form of the histogram if you split your sample into two equal groups, the first close to the galactic plane, and the second further away.

TREFZGER: (1) The errors in our determinations of (B-V) are rather small, about 0.03 - 0.04 mag. (2) We have not yet looked into the behavior of (B-V) as a function of distance.

THE VERTICAL DISTRIBUTION OF STARS: THICK DISK OR NO THICK DISK?

Annie C. Robin, Michel Crézé, Olivier Bienaymé,
Edouard Oblak

Observatoire de Besançon

ABSTRACT: Previous investigations of galactic structure, based on either observations of distant giants or from general star counts, have given some evidence that an intermediate population could be necessary for a realistic description of the Galaxy, while some other investigations do not find any evidence for it. We try here to find new constraints on this population using a sample at intermediate latitude which includes photometry and proper motions. We show that a proper statistical analysis of the data in the 4-dimensional space (V, B-V, μ_l, μ_b) in comparison with a model of population synthesis allows one to constrain the thick disk distribution. Although this sample does not permit us to distinguish between intermediate populations of scale height between 800 and 1200 pc, there is good evidence that a thick disk is necessary to fit the distribution of observable quantities in the 4-dimensional space. This result rules out the possibility that thick disk stars belong to the tail of the distribution of the old disk. It implies some kind of discontinuity between disk and thick disk in the scenario of star formation and/or dynamical heating.

1. INTRODUCTION

The existence of an intermediate stellar population, call it a thick disk, between the disk and halo is a controversial matter. Different approaches have been used, based on either observations of distant giants, where giants with relatively high metallicities and large rotational velocities relative to halo stars are found (Rose 1985, Ratnatunga and Freeman 1985, 1989), or star counts (Gilmore and Reid 1983, Wyse and Gilmore 1986, Yoshii et al. 1987). Parameters describing this population are a scale height ranging from 600 to 1500 pc, local density (1% to 12% of the disk), velocity ellipsoid (with a vertical velocity dispersion from 30 to 80 km/s) and an asymmetric drift (20 to 100 km/s), while some other investigations have not found any evidence for a thick disk. Freeman (1987) reviewed the main contributions to the subject. Apart from evidence of thick disks in external spiral galaxies (Van der Kruit and Searle 1981, 1982)

A. G. Davis Philip and P. K. Lu (eds.)
The Gravitational Force Perpendicular to
the Galactic Plane   33 - 43
©   1989   L. Davis Press

different values of the parameters determined for the thick disk of the
Milky Way are summarized in Table I.

Samples of distant giants as tracers of this population suffer
from their small numbers. On the other hand general star counts in
magnitude and color fail to distinguish a 2-component from a
3-component galaxy. Proper motions are expected to make a serious
contribution to the evidence for this population, which is supposed to
have distinct characteristics in velocities compared to either the disk
population or the halo. A previous investigation of Wyse and Gilmore
(1986) using Chiu's survey of stellar magnitudes, colors and proper
motions showed the probable existence of a thick disk but the parameter
determination was poor since the observations covered a very small area
of sky (3 times 0.1 square degrees) with about 500 stars of magnitudes
between 12 to 20, a small number of which contributed to the thick
disk. The information coming from the colors was not used.

We propose here to qualify the thick disk using observations in a
4 parameter space (V, B-V, μ_l, μ_b) at intermediate galactic latitudes
and a suitable model of population synthesis. In section 2 we briefly
recall the galaxy model. In section 3 we try to identify the
intermediate population using colors and proper motions. In section 4
we apply a statistical analysis to compare data with models having
different shapes and velocity ellipsoids for the thick disk.

2. MODELING THE GALAXY AND ITS KINEMATICS

The mainsprings of the model of population synthesis have been
described in Robin and Crézé (1986) and form the basis for stellar
evolution and the photometry. Bienaymé et al. (1987) have described
the overall dynamical consistency, and Robin and Oblak (1988) have
discussed the kinematics. The main ideas have been recalled in Crézé
(1989). It is useful to recall here that the model allows one to
produce for any galactic direction either suitable statistics of
observable quantities and intrinsic properties of stars, or a catalogue
of pseudo-stars in which all stellar properties are stored. Observable
quantities include apparent magnitudes in a chosen system, colors,
proper motions and radial velocities. The intrinsic properties of
stars are described by the absolute magnitude, spectral type, age and
the three velocities U, V and W.

3. SEARCH FOR SIGNATURES IN THE (V, B-V, μ_l, μ_b) SPACE AT INTERMEDIATE
LATITUDES

At intermediate latitudes thick disk stars are expected at
magnitudes fainter than V = 16. To check we produced model predictions
in the direction of M 5 (l = 2.7°, b = 45.3°) where observations exist.
The predicted catalogue can be used to identify the portion of the
4-dimensional space where the thick disk has distinct observable
properties compared to the disk and the halo. Fig. 1 shows the

TABLE I

Summary of Some Previous Determinations of the Intermediate
Population Parameters

Authors	Method	Metallicity	σ_w km/s	Rot. lag km/s	Scale Height pc
Blaauw (1965)	Compilation	[Fe/H] -.4	25		700
Janes (1975)	Nearby giants	δCN <-0.08	35 ±6	35	
Hartkopf-Yoss (1982)	Distant giants	[Fe/H] > -1	39 ±6		
Gilmore-Reid (1983)	Photometric counts				1500
Rose (1985)	Red horiz. branch	[Fe/H] -.7	40		
Norris (1986)	Distant giants		70 ±10*	60 ±30	
Wyse-Gilmore (1986)	Proper motions		60	100	1500
Yoshii et al. (1987)	Photometric counts				950
Ratnatunga-Freeman (1989)	Distant giants	[Fe/H]>-.8	50 ±10	36 ±14	

* Line of sight velocity dispersion

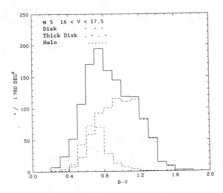

Fig. 1. Predicted (B-V) distribution of stars towards M 5 (l =
2.7°, b = 47.3°) in the magnitude
range 16 < V < 17.5 in a 1.78
square degree region. The
assumed measuring error in (B-V)
is 0.1 mag. Full line - all stars;
dashed line - disk; dotted-dashed -
thick disk; dotted line - halo.

Fig. 2. Same as Fig. 1 but for
stars with positive μ_l. Full line
- disk; dotted-dashed line - thick
disk; dotted line - halo.

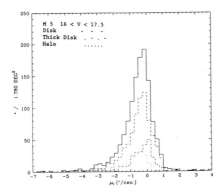

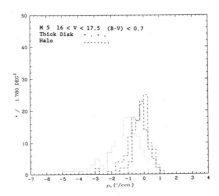

Fig. 3. Predicted μ_l distribution
of stars towards M 5 (l = 2.7°, b =
47.3°) in the magnitude range 16 <
V < 17.5 in a 1.78 square degree
region. The assumed measuring
error in the proper motions is 0.2
"/century. Full line - all stars;
dashed line - disk; dotted-dashed -
thick disk; dotted line - halo.

Fig. 4. Same as Fig. 2 but for
stars of (B-V) < 0.7 only.
Dashed line - disk; dotted-dashed
- thick disk; dotted line - halo.

distribution of stars in (B-V) in this direction (solid line) and the distribution of each population (dashed line - disk, dotted line - halo, dashed-dotted - intermediate population) predicted by the model in the magnitude interval 16 to 17.5. It shows that the colors contain information about the three populations but they are not very well separated. Fig. 2 shows the same distribution for stars with positive μ_l only. The separation between thick disk and halo stars is much easier since very few halo stars have positive μ_l because of their large V velocity with respect to the LSR. Moreover the contrast between disk and thick disk is higher. Fig. 3 shows the predicted global distribution of stars in μ_l and the separated populations as in Fig. 1. The μ_l proper motion is parallel to the V velocity in this direction such that the three populations can be separated using their rotational lag as well as their velocity dispersions. Fig. 4 shows the same distribution as Fig. 3 but for stars of (B-V) < 0.7 only. Again the contrast between the populations is enhanced by using a color index.

These examples help in understanding the respective positions of the different populations in the (V, B-V, μ_l, μ_b) space but we lose information using only 2-D projections. A proper statistical analysis can be used instead to compare the global 4-dimensional distribution of observed stars with model predictions.

4. THE STATISTICAL METHOD

The comparison between data and model predictions is made by binning the magnitudes V, colors (B-V) and proper motions to obtain a discrete distribution function N (V, B-V, μ_l, μ_b). In the case of model predicted data we count separately stars of the three populations and separate stars by distances. The corresponding function of the model is $\Sigma\, M_i$ (V, B-V, μ_l, μ_b) where i is an index which differentiates populations and distance ranges. In the analysis the model stars have been grouped into two bins of distances (r < 1 kpc, and r > 1 kpc, which is very close to the median of the distribution in distances) and two or three population bins (disk, thick disk when emphasized, halo). The relation between i and the populations is given in the first and second column of Table III. For example i = 1 represents stars of the disk population closer than 1 kpc.

The separation of model predictions in population and distance bins allows one to constrain the proportion of each population that gives the best agreement with the data. We resolve a system of equations where we impose in each bin a linear combination of the model counts to be equal to the data counts. That is

$$N = \Sigma\, C_i M_i \qquad (1)$$

TABLE II

Characteristics of each tested model. Densities are in stars
per cubic parsec and velocity dispersions in km/s.

Model	Old disk vel. disp	Halo density	Thick disk vel. disp	Thick disk Scale Ht.
2-component models (Disk, halo)				
(0) standard disk	(47,30,25)	1.93 E-4		
(1) thicker old disk	(51,33,30)	1.93 E-4		
(2) thicker old disk	(60,39,35)	1.93 E-4		
(3) Std Disk, halo *1.5	(47,30,25)	2.90 E-4		
3-component models (Disk, Thick disk and halo)				
(4)	(47,30,25)	1.93 E-4	(80,55,50)	1170
(5)	(47,30,25)	1.93 E-4	(60,39,35)	500
(6)	(47,30,25)	1.93 E-4	(75,50,45)	800
(7)	(47,30,25)	1.93 E-4	(100,77,70)	1500

TABLE III

Statistical Estimation of the Different Models.

i	Population		(0)	(1)	(2)	Model (3)	(4)	(5)	(6)	(7)
1	Disk r< 1kpc	N=	375	410	425	384	375	376	388	378
		C_i=	0.87	0.88	0.95	0.83	0.62	0.62	0.62	0.61
		$\sigma(C_i)$=	0.08	0.08	0.08	0.09	0.10	0.10	0.10	0.09
2	Disk r> 1 kpc	N=	148	212	278	155	148	156	155	157
		C_i=	2.10	1.52	1.39	2.14	2.08	1.08	1.47	1.72
		$\sigma(C_i)$=	0.11	0.09	0.08	0.12	0.17	0.24	0.21	0.17
3	Halo r< 1 kpc	N=	0.6	0.7	0.5	1.3	0.6	0.5	0.5	0.6
4	Halo r> 1 kpc	N=	200	203	198	309	200	205	202	207
		C_i=	1.6	1.45	1.17	0.92	1.74	1.54	1.45	1.41
		$\sigma(C_i)$=	0.13	0.14	0.14	0.09	0.16	0.13	0.16	0.19
5	I.P. r < 1 kpc	N=	0	0	0	0	15	9	13	12
6	I.P r > 1 kpc	N=	0	0	0	0	208	31	99	313
		C_i=					-0.09	5.29	1.43	0.31
		$\sigma(C_i)$=					0.21	1.51	0.49	0.17
	Likelihood		6.5	4	4.	7.5	10.	6.	5.	8.
	S.D. plain model		3.97	3.73	3.51	3.79	3.22	3.68	3.44	3.48
	S.D. modified model		2.97	3.31	3.25	3.26	2.92	3.01	2.99	3.22

r is the heliocentric distance. I. P. is the intermediate population or thick disk. N is the number of stars of the i contributor of the model. C_i is the computed coefficient of eq. (1) of the i contributor and $\sigma(C_i)$ is its rms error. The coefficient of the halo and thick disk stars closer than 1 kpc are not shown because so few stars are concerned that C_i has no meaning. S. D. means standard deviation. The likelihood estimator is given in number of sigmas and refers to C_i coefficients.

Coefficients C_i and their accuracy are determined by least square fit
to each bin of the data space (V, B-V, μ_l, μ_b). The large number of
condition equations (several hundred) allows us to determine up to 10
coefficients and their rms errors. The number of different model
contributors in equation (1) is limited so that the total number of
stars in each group is large enough to avoid poor statistics. The
model is compatible with the data when all coefficients are equal to
unity within three sigmas. If the accuracy of a given coefficient is of
the order of one, no information on this coefficient can be found from
this set of data. Coefficients can be correlated with each other.
Such results cannot be taken into account independently. Some tests
have been made to be sure that the size of the bins used in the N (V,
B-V, μ_l, μ_b) distribution does not influence the result. This can
happen if the number of stars in each bin is too small to give any
information. The optimal size of the bins we used is 1.0 magnitude in
V, 0.4 magnitude in (B-V) and 1"/century in proper motion.

This method allows us to estimate: 1) the adequacy of the model
to reproduce the data, 2) the corrections in the form of linear
coefficients to apply to the model to better reproduce the data, 3) the
adequacy of the modified model to reproduce the data, 4) the
significance level with which the coefficients have been determined.
The accuracy of the method is limited by the use of discrete functions
rather than smooth distributions and may be sensitive to systematic
errors in the data.

We applied the method to a set of eight different models. First,
a model with no thick disk at all (standard model with intermediate
population removed) was tried. Then, two models with no thick disk but
where the oldest disk stars have larger velocity dispersions and
accordingly higher scale height were tried. Then, a model with higher
halo density, as if thick disk stars could be replaced by halo stars,
was tried. Finally, four models were tried with a separated
intermediate population with different scale heights and velocity
ellipsoids. The physical parameters of each model are given in Table
II.

In Table III we give the result of the least squares fit of each
model to the data: computed coefficients C_i and their rms accuracy
$\sigma(C_i)$, number of stars concerned N, the global standard deviation of
the plain model i.e. with all C_i coefficients equal to 1 (it measures
the adequacy of the emphasized model), the standard error of the
modified model with the new coefficients (a measure of the adequacy of
any model with same velocity ellipsoids).

We also compute the likelihood measuring the level of
significance of the computed coefficients. We give here the relative
value of the likelihood in the value of sigmas. If this likelihood is
less than three sigmas, the model is compatible with the data but not

strongly constrained by it. All coefficients are compatible with one sigma (according to their rms error).

5. CONCLUSIONS

We may draw the following conclusions from Table III. The models involving a thicker old disk (models (1) and (2)) are unlikely in view of their global standard deviation even after adjustment of the coefficients. It means that a part of the 4-D space is not well represented by this model and this cannot be changed just by changing the proportion of stars in the different populations and distance bins.

The model (0) with a normal disk and halo but no thick disk gives good predictions for nearby disk stars but is in disagreement with the number of distant disk stars by a factor of two. However the modified model found in this way gives a good standard deviation compared to the other models. It means that stars lacking at $r > 1$ kpc could have velocities and scale heights similar to the old disk population.

The model with a halo with higher density does not give a good global fit, even modified, although it reinforces the idea that the halo density emphasized in the standard model may be too low. This conclusion should be verified using data obtained in complementary directions.

It should be noted that the correlations between coefficients obtained for the nearby thick disk and the nearby disk, on the one hand, and the farther thick disk and the halo, on the other hand, are sometimes a bit high (about 0.5 to 0.6 for the first one, and 0.6 for the second one). However, when combined, these coefficients are quite stable from one model to another. For example the halo density is found to be too low by 50% or so in all models with the standard normalization.

The different models with a true intermediate population give the best agreement with the data either in their plain form or in their modified form. The model with a thick disk of 1500 pc scale height is unlikely and our best model is the one with a scale height of 800 pc. But all models with scale heights from 500 pc to 1200 pc and vertical velocity dispersions between 35 and 55 km/s give an acceptable fit. Complementary data in other directions would be useful to better constrain these parameters.

The small number of thick disk stars relative to the old disk (between 1% and 2% locally) and the somewhat larger velocity dispersions makes it improbable that these stars belong to the tail of the old disk distribution, as we can see by comparing the data with models involving large dispersions for old disk stars. For reasons of continuity, it is difficult to imagine a physical heating process for the disk working only on a small proportion of old disk stars. A scenario with separated formation for the old disk and the intermediate population would be more realistic.

REFERENCES

Bienaymé, O., Robin, A. C. and Crézé, M. 1987 Astron. Astrophys.
    180, 94.
Blaauw, A. 1965 in Galactic Structure, A. Blaauw and M. Schmidt,
    ed., University of Chicago Press, Chicago, p. 435.
Crézé, M., Robin, A. C. and Bienaymé, O. 1989 in
    The Gravitational Force Perpendicular to the Galactic Plane, A.
    G. D. Philip and P. K. Lu, eds., L. Davis Press, p. 3.
Freeman, K. C. 1987 Ann. Rev. Astron. Astrophys. 25, 603.
Gilmore, G. and Reid, N. 1983 Mon. Not. Roy. Astron. Soc.
    202, 1025.
Hartkopf, W. and Yoss, K. M. 1982 Astron. J. 87, 1679.
Janes, K. A. 1975 Astrophys. J. 29, 161.
Norris, J. 1986 Astrophys. J. Suppl. 61, 667.
Ratnatunga, K. U. and Freeman, K. C. 1985 Astrophys. J. 291,
    260.
Ratnatunga, K. U. and Freeman, K. C. 1989 Astrophys. J. 339,
    126.
Robin, A. C.and Crézé, M. 1986 Astron. Astrophys. 157, 71.
Robin, A. C. and Oblak, E. 1987 Proc. 10th IAU European Astronomy
    Meeting, Vol. 4, J. Palous, ed., p. 323.
Rose, J. 1985 Astron. J. 90, 787.
Van der Kruit, P. C. and Searle, L. 1981 Astron. Astrophys.
    95, 105.
Van der Kruit, P. C. and Searle, L. 1982 Astron. Astrophys.
    110, 61.
Wyse, R. F. G. and Gilmore, G. 1986 Astron. J. 91, 855.
Yoshii, Y., Ishida, K. and Stobie, R. S. 1987 Astron. J. 92,
    323.

DISCUSSION

SCHECTER: How important is the inclusion of proper motions?

ROBIN: You can hardly separate thick disk from other populations from
magnitude and colors alone. Two component models generally fit star
counts quite well. But the different velocity dispersions of the
different populations give an opportunity to separate them by proper
motions.

GOULD: What is the rough scale of your astrometry errors in terms of
velocities?

ROBIN: An accuracy of 0.2" per century in proper motion gives an error
in velocity of about 10 km/s at 1 kpc.

SCHECHTER: I'd like to ask John Norris what the prospects are for discriminating between two discrete components and one continuously varying component.

NORRIS: I feel that unless you have components having a higher velocity dispersion than 35 km/sec in your thick, old disk model you won't get much improvement.   What would happen, for example, if you used 50 km/s for your hotter component; or better still, included a continuous distribution up to this limit?

ROBIN: I have not used a component that thick.   I thought it was unrealistic to put a very populated component (the old disk) to so high scale height, but I will try.   Another problem will arise then.   The potential will be modified quite a bit with such a distribution.

GILMORE: 1) Have you analyzed Murray's SGP proper motion catalogue? 2) Since your best-fit model still is a three sigma residual, does this mean that none of the models you tried is acceptable?

ROBIN: 1) I have analyzed Murray's data (Robin and Oblak, 1987) and the conclusion is that their proper motions are not accurate enough to get any detailed information on the thick disk.   Their time scale is too short (about 2 years).   2) All the models are about at three sigma. The reason is that the systematic errors are quite important when you put stars into bins.   I guess that the main systematic errors come from photometry.   A better statistical method is needed to solve this problem.

TREFZGER: On the basis of your models, can you predict the metallicity distributions at different z values in the directions of the galactic poles so as to discriminate between various models of the thick disk?

ROBIN: Yes, we can produce metallicity distributions.   It would be very interesting to complete our velocity constraints with metallicity from your data to get a better discrimination of the thick disk.

RATNATUNGA: When you compare the observed distribution with the predicted distribution do you first average a number of simulations or compare a number of simulations?

ROBIN: Predicted distributions are obtained for ten times the area of the observed data equivalent to a mean of ten simulations.

A PROPER MOTION STUDY IN SA 57

A. Spaenhauer

Astronomical Institute, Basel

ABSTRACT: We present some results based on a medium deep proper motion (down to V ~ 17 mag) and photographic B, V (down to V ~ 18 mag) survey in the North Galactic Pole field SA 57. Using statistical photometric parallaxes for a subsample of F- to early G-type stars the photometric data are in perfect agreement with a Galaxy model including a thick disk component with a scale height of ~ 1300 pc. The kinematic data indicate a rotational lag of the thick disk relative to the LSR of ~ 60 to 80 km/s and V velocity dispersion of ~ 50 km/s. The U and V velocity dispersions $(\sigma_v/\sigma_u)^2 = 0.43$ of the disk stars at ~ 1 scale height above the plane (~ 300 pc) are in remarkable agreement with the rotational constants derived from local solar neighborhood stars.

1. INTRODUCTION

Since the papers of Bahcall and Soneira (1981) and Gilmore and Reid (1983) Galaxy modeling has become a major tool for studying the global features of the Galaxy. Following the ideas of Kapteyn, enormous progress has been made in acquiring data in selected fields of the Galaxy. The top-down philosophy of the modeling idea (model → actual predictions → comparison with observations → model correction) is methodologically clear but needs, in the case of a galaxy, many input parameters which are taken from extragalactic observations (e.g., overall density laws) and local galactic studies (normalizations). One of the topics rediscussed in recent years (see e.g., Blaauw and Schmidt 1965) is the question whether the observed data require an intermediate field star component in addition to an exponential disk with scale height of ~ 300 pc and a spheroid. Recently much data have been presented advocating the existence of such a component. One of the remarkable points is that the existence of such a component (thick disk) follows not only from photometric data (see e.g., Fenkart 1988) but also from kinematic data (Sandage 1987, Norris 1986, Freeman 1987, Wyse and Gilmore 1986). In this study new proper motion data in conjunction with photometric data in a field close to the North Galactic Pole, SA 57, are used to derive some properties of the z-structure of the Galaxy

A. G. Davis Philip and P. K. Lu (eds.)
The Gravitational Force Perpendicular to
the Galactic Plane   45 - 52
©   1989   L. Davis Press

2. DATA

The available data consist of complete photographic B and V photometry down to V = 18 mag and proper motions of all the stars down to V = 17 mag in the 2.6 square degree field of SA 57 with coordinates l = 65.5° and b = 85.5°. The photographic B and V magnitudes have been determined from six B plates and six V plates taken with the 48" Palomar Schmidt telescope. The measurements and reductions have been carried out using an iris photometer and using the photoelectric sequence given by Purgathofer (1969). The total mean errors of the (B-V) colors are estimated from a comparison with Weistrop's data (Weistrop 1972) using the corrections given by Faber et al. (1976) and are ~ 0.05 mag. The photometric data are published in the Basel Photometric Catalogue Nr. 1 and a discussion of these photometric data is published in Fenkart (1967) and Fenkart and Esin-Yilmaz (1985). The proper motions have been determined using a total of six Lick astrograph plates (scale = 55"/mm) well distributed over a baseline of 36.6 years (1942.45 to 1979.08). All the plates have been measured with the automatic measuring engine (AME) at Lick Observatory in direct and reverse mode in order to avoid any instrumental magnitude terms. The plate to plate reductions were carried out using the central overlap method with 10 plate constants described by Eichhorn (1970). 30 well measurable galaxies in the field could be used as a reference frame defining the zero point for the resulting absolute proper motions.   The errors of this zero point mean motion of the galaxies) are 0.11"/100 y and 0.13"/100 y for μ_α and μ_δ respectively. The total random errors of the absolute proper motion in the galactic coordinate system are shown in Fig. 1 as a function of the apparent V magnitude. Because our field lies close to the North Galactic Pole, the measured proper motion can be directly converted into proper motion in the galactic system by a simple rotation:

$$U= 4.74.R.\mu(l=0^0) = 4.74.R.(0.82.\mu_\alpha - 0.57.\mu_\delta) \qquad (+0.03W)$$
$$V = 4.74.R.\mu(l=90^0) = 4.74.R.(0.57.\mu_\alpha + 0.82.\mu_\delta) \qquad (+0.07W)$$

U, V and W denote the velocities in the galactic coordinate system relative to the Sun (toward the galactic center, in the direction of galactic rotation and perpendicular to the galactic plane respectively) and R is the distance to the star. The terms in parentheses denote the errors caused by the fact that the galactic latitude is not exactly 90°. These errors are 0.03 W and 0.07 W and will be neglected in the following.

3. RESULTS

We want to present some results based on a subsample of the SA 57 stars.   To transform the absolute proper motion into absolute velocities relative to the Sun, we have to know the (photometric) distances of the stars. Having (B - V) colors only, we do not know the evolutionary status of the stars and hence we decided to select a subsample of the stars based on their colors. The color regime 0.3 mag

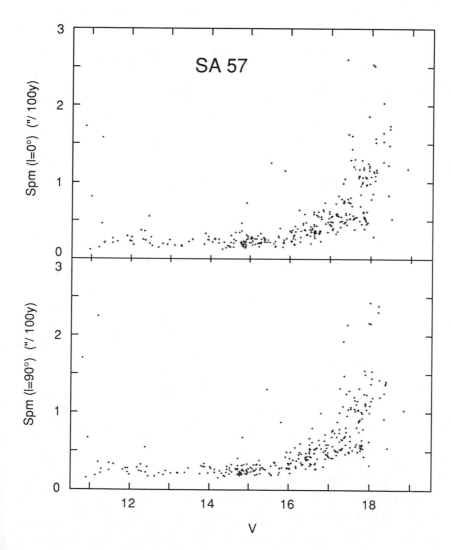

Fig. 1. Random errors of the proper motion components (in the galactic coordinate system) as a function of the V magnitudes.

$\leq$ (B - V) $\leq$ 0.6 mag (F- to early G-type stars) seems to be best suited for our purpose because it avoids the giant main sequence bimodal absolute magnitude distribution that appears in the redder vertical strips of the color-magnitude diagram. Furthermore these stars are the brightest and hence farthest reaching main sequence stars in this direction. The mean absolute magnitude of these stars we assume to be 4.5 mag. The overall density as a function of distance (height above the plane) is shown in Fig. 2 using the complete photometry down to V = 18 mag. Also shown is the decomposition of the observed density into three components: thin disk with a scale height of 300 pc, thick disk with a scale height of 1300 pc and a spheroidal component with a local scale height of about 4500 pc (Gilmore and Wyse 1985). No attempt has been made to derive a "best" decomposition because the parameter space is too large for our broadband photometric data. The relative normalizations follow naturally from the adopted scale heights mentioned above and assuming that the thin disk dominates up to ~ 500 pc and that the thick disk dominates around 2000 pc:

$$\rho_{o,thin\ disk} : \rho_{o,thick\ disk} : \rho_{o,spheroid} = 700 : 15 : 1$$

A decomposition of the observed density function into the two classical components thin disk and spheroid yields a significant lack of stars in the range two to three kpc as well as a density normalization at the solar neighborhood of ~ 300 : 1.

In the following we want to extract some information about the kinematic z - structure of the F- to early G - type stars. All the subsequent analysis is based on the assumption that the absolute magnitudes of the stars are 4.5 mag. Calculations have been carried out using 4.0 mag and 5.0 mag but they do not change the conclusions essentially for the stars up to ~ 1 kpc. The kinematic parameters for the thick disk are changed by ~ 20%. From the color selection described above (302 stars) we excluded for this study all the stars with proper motion errors $\geq$ 0.07"/100 y. This proper motion error limit transforms to a velocity error limit of ~ 24 km/s per kpc. A total of 219 stars remain in the sample (191 up to a distance of 3500 pc). Table I summarizes the statistics of the sample for different distance intervals (heights above the plane). Fig. 3 shows the mean U and V velocities as a function of height above the plane. The dotted line shows the solar motion relative to the local standard of rest ($<U>$ = -9 km/s, $<V>$ = -12 km/s). Whereas Fig. 3a shows no systematic radial motions of the stars up to ~ 3 kpc within the errors, Fig. 3b shows a steadily increasing rotational lag relative to the Sun and LSR. This behavior is expected and reflects the varying contributions of thin disk, thick disk and spheroid stars to the observed mean rotational lag. Close to the Sun (within ~ 1 kpc) almost only disk stars are observed rotating with about solar speed whereas towards the limiting distance of ~ 3 kpc thick disk and spheroid stars dominate showing a considerable rotational lag relative to the Sun. Obviously we do not reach a pure spheroidal population with our limiting magnitude of ~17

TABLE I

Statistics of the F- to Early G-Type Stars towards the North
Galactic Pole (2.6 sq. degrees) grouped into distance intervals.

distance interval (pc)	0 - 500	500 - 1000	1000-1500	1500-2500	2500-3500
Number of stars	25	23	52	46	45
<U>	1 km/s ±18 km/s	-17 km/s ±10 km/s	-16 km/s ±14 km/s	-24 km/s ±13 km/s	+14 km/s ±22 km/s
<V>	-4 km/s ±6 km/s	-15 km/s ±10 km/s	-54 km/s ±10 km/s	-95 km/s ±12km/s	-148 km/s ±18 km/s
σ_u	42 km/s	50 km/s	99 km/s	89 km/s	145 km/s
σ_v	28 km/s	46 km/s	72 km/s	80 km/s	122 km/s
mean error of velocity comp.	4 km/s	9 km/s	16 km/s	34 km/s	67 km/s
<distance>	330 pc	790 pc	1230 pc	2050 pc	2900 pc

Meaning of the kinematic symbols: <U> is the mean U velocity component
(towards the galactic center) and its error.   <V> is the mean V
velocity component (direction of galactic rotation) and its error. σ_u
is the dispersion of the U velocities. σ_v is the dispersion of the V
velocities.

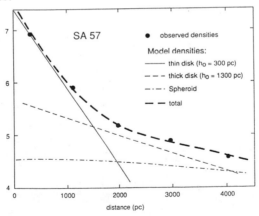

Fig. 2. Logarithmic density gradient of the F- to early G-type stars
in the direction of the North Galactic Pole (SA 57). Also shown is the
decomposition into three components according to a Gilmore - Wyse
model.

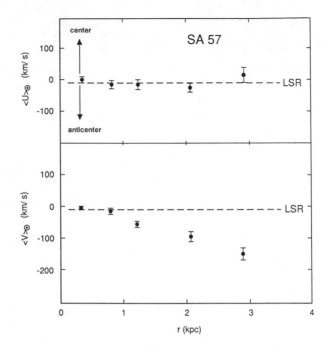

Fig. 3.   The transverse mean velocity components U (relative to the Sun) and V (relative to the Sun and in the direction of galactic rotation) as a function of distance (height above the plane).

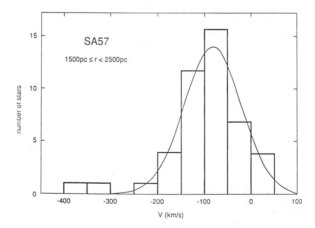

Fig. 4.   Histogram of the V velocities (relative to the Sun) of the stars in the distance interval 1500 pc to 2500 pc (mainly thick disk stars).   The smooth curve is a Gaussian fit (excluding the two high velocity stars) with a mean at 85 km/s and a dispersion of 60 km/s.

mag. Up to 500 pc above the plane we have practically a pure thin disk population and we can therefore obtain information on Oort's rotational constants of disk stars at ~ 1 scale height above the plane. From Table I we derive $(\sigma_v/\sigma_u)^2 = 0.43$ for stars with a mean height above the plane of 330 pc. This value is in remarkable agreement with the values of A = 14.5 km/s/kpc and B = -12 km/s/kpc derived from solar neighborhood stars and using the formula $-B/(A-B) = (\sigma_v/\sigma_u)^2$. Between 1500 pc and 2500 pc above the plane, the thick disk dominates. Theoretically the relative star numbers in this distance interval should be according to a three component model ~ 20 : 70 : 10 (thin disk : thick disk : spheroid). However, because of our kinematic proper motion error cutoff and the small number of stars (46) we do not try to disentangle these three components (especially thin and thick disk) in the histogram of the V velocities shown in Fig. 4. An overall Gaussian fit (excluding only the two high velocity stars with velocities between -300 km/s and -400 km/s) yields a mean rotational lag relative to the LSR of ~ 60 km/s to 80 km/s and a dispersion of ~ 50 km/s (after deconvolution with the observational errors of ~ 35 km/s). Assuming systematically brighter absolute magnitudes (~ 4 mag) would increase these values by about 10 km/s to 20 km/s.

4. FUTURE WORK

     The results described above are based on broadband photometric parallaxes and their interpretation is, of course, model dependent. The conclusions drawn above are greatly insensitive to the specific proportions of absolute magnitudes among the turn-off stars but they are sensitive to an eventual contamination by red horizontal-branch (RHB) stars. However, due to our color selection (0.3 mag $\leq$ (B - V) $\leq$ 0.6 mag), we believe that no significant fraction of RHB stars is expected. In order to arrive at more precise conclusions about the parameters of the different Galaxy components (including metallicity information), intermediate band photometry giving metallicities and distances and/or spectroscopic data giving metallicities and radial velocities of the F- to early G-type stars are necessary. A program has been initiated at Calar Alto Observatory (by Spaenhauer and del Rio) of getting, in a first phase, uvby photometry of all the F- to early G-type stars in SA 57 down to ~ 15 mag.

ACKNOWLEDGEMENTS

     I would like to thank B. Jones and A. Klemola for their invaluable help during my stay at Lick Observatory and for their willingness to let me use the AME for the proper motion measurements as well as the Swiss National Science Foundation for financial support.

REFERENCES

Bahcall, J. N. and Soneira, R. M. 1980 Astrophys. J. Suppl.
     44, 73.
Blaauw, A. and Schmidt, M., eds. 1965 Galactic Structure

University of Chicago Press, Chicago.
Eichhorn, H. 1974 in Astronomy of Star Positions, Frederic Ungar
   Publishing Co., New York.
Faber, S. M., Burstein, D., Tinsley, B. M. and King, I. R. 1976
   Astron. J. 81, 45.
Fenkart, R. P. 1967 Z. Astrophys. 66, 390.
Fenkart, R. P. and Esin Yilmaz, F. 1985 Astron. Astrophys. Suppl.
   62, 39.
Fenkart, R. P. 1988 Astron. Astrophys. Suppl. 76, 469.
Freeman, K. C. 1987 Ann. Rev. Astron. Astrophys. 25, 603.
Gilmore, G. and Reid, N. 1983 Monthly Not. Roy. Astron. Soc.
   202, 1025.
Gilmore, G. and Wyse, R. F. G. 1985 Astron. J. 90, 2015.
Norris, J. 1986 Astrophys. J. Suppl. 61, 667.
Purgathofer, A. Th. 1969 Lowell Obs. Bull. No.147, VII, 98.
Sandage, A. 1987 in The Galaxy, NATO ASI Series, G. Gilmore and
   B. Carswell, eds., Reidel, p. 321.
Weistrop, D. 1972 Astron. J. 77, 849.
Wyse, R. F. G. and Gilmore, G. 1986 Astron. J. 91, 855.

DISCUSSION

DA COSTA: How does the mean absolute magnitude for the stars with Strömgren photometry agree with the value assumed for the analysis of the broad band data? What is the dispersion in M_v values?

SPAENHAUER: For the Strömgren sample, which is about complete to V = 14.5 mag, the mean absolute V-mag. is 4.3 mag with a dispersion of about 0.5 mag.

FREEMAN: In your interval z = 1500 to 2500 pc, in which you found an asymmetric drift of about 70 km/s (relative to the LSR), how much contamination do you estimate from the (slowly rotating) halo population?

SPAENHAUER: According to a Gilmore-Wyse three component galaxy model the relative contributions in this distance interval are about 70% (thick disk), 20% (thin disk), 10% (Halo). Therefore, the contribution from the halo should be negligible.

KING: This may seem a minor technical point, but I think it can be quite important when trying to discriminate between continuous and discrete transitions. In taking your statistics, the median would be much less influenced by a small contaminating population than the mean is influenced. Similarly, the quartile would be much less influenced than the root-mean-square.

SPAENHAUER: I agree with that statement. The fit in Fig. 4 has been done excluding the two halo stars so that the mean should roughly agree with the median.

MAGNITUDE, COLOR AND PROPER MOTION SURVEY IN A DIRECTION TOWARDS M 5

O. Bienaymé, V. Mohan, M. Crézé, S. Considère and A. Robin

Observatoire de Besançon

ABSTRACT: We present a new magnitude-color-proper motion survey in the direction $l = 2.7°$, $b = 47.3°$. The survey covers 1.78 square degrees near the globular cluster M 5. It is complete in UBV to V = 16.5 and in BV to V = 17.5. The accuracy in relative proper motions is 0.2"/century; they are computed on a time baseline of 31 years. We describe a few relevant points of the astrometric procedure.

1. INTRODUCTION

Star counts are used as tracers of density distributions in the Galaxy. Multicolor photometry of the survey stars can give insight into the luminosity function and temperature distribution of the stars. Proper motions contain kinematic information together with distance information directly related to the kinematics of the Galaxy and can be obtained for a large sample of stars in the survey. Radial velocities associated with the deep photometric survey are measured only for selected samples; the amount of observational time necessary to measure radial velocities for a few thousand stars up to 18th magnitude remains out of reach. In star count surveys, intrinsic properties of individual stars such as the absolute magnitude, the metallicity, the luminosity class or the distance remain unknown. A fruitful analysis method developed by many authors (Ratnatunga et al. 1989, Robin and Oblak 1987) consists of building a galactic model giving predictions that are directly compared to the observational data. This modeling approach directly links physical quantities to the distributions of observable quantities rather than trying to recover intrinsic properties of individual stars from broad-band photometry and proper motions since no one-to-one link does exist between this kind of observable and intrinsic properties. The main model inputs are the density distributions, the luminosity function and the kinematics of stars (available from local observations). These quantities can be constrained by theoretical considerations (using Boltzmann and Poisson equations) to ensure dynamical consistency at the solar radius towards the galactic poles.

Magnitude and proper motion data obtained for the same stars allow us to consider the information available in a multidimensional

A. G. Davis Philip and P. K. Lu (eds.)
The Gravitational Force Perpendicular to
the Galactic Plane   53 - 59
©   1989   L. Davis Press

Combined data on the same sample give direct insight into the phase space distribution. While star counts give only the total density distribution, proper motion surveys at high galactic latitude give access to the distribution of each isothermal component. The existence of a thick disk, separated or not from the old disk, are hypotheses that would be revealed by a signature in the proper motion distributions (Robin et al. 1989). In the case of another (proper motion-color-magnitude) survey in progress towards M 37 (anticenter), information on the luminosity function as well as the star formation rate could be extracted from the data (Robin et al. 1989).

2. PROPER MOTION SURVEYS

Such sample surveys should match a few quality conditions in order to give new constraints on models and new insight into galactic structure. The survey area should be large enough to allow a convenient statistical analysis. The accuracy in the photographic photometry is expected to be about 0.05 magnitude but the difficulties in reduction and calibration and the inconsistency between different surveys suggest that slightly higher systematic errors could exist. The proper motion accuracy is related to the time baseline. Two recent proper motion surveys with accurate photometry are in fact limited by their short time baseline or small limiting magnitude or small analyzed area. The Murray et al. (1986) catalogue, based on Schmidt plates, covers an area of 20 square degrees towards the South Galactic Pole in B and V. It is complete to V = 16. However the proper motion accuracy is 1.4"/century since the time baseline is two years. The Chiu (1980) catalogues, based on prime focus plates (towards three directions), cover very small areas of 0.1 square degree with BV complete to V = 20.5. The accuracy in relative proper motions is 0.04"/century since the time baseline is 20 to 25 years (for absolute proper motion based on background galaxies the accuracy is 0.2"/century). So none of these catalogues can give new kinematic constraints (Robin and Oblak 1987). To go further we need a proper motion catalogue covering a larger area with a larger time baseline.

3. "M 5" SURVEY

We attempted to measure proper motions and photometry using Schmidt plates separated by 31 years (a few plates were taken from 1981 to 86 and one is a Palomar Observatory Sky Survey copy of a plate taken in 1955). The area covered by our survey is 1.78 square degrees and is complete in BV to V = 17.5 and in UBV to V = 16.5. We did not make an astrometric plate solution using reference stars for astrometric solution and measured only relative proper motions. We obtained an accuracy of 0.2"/century.

3.1 Photometry

The photometry was done with Schmidt plates in U(1), B(2), V(2). Plates were taken from 1981 to 1986 with the Observatoire de la Côte d'Azur Schmidt telescope.   The plate to plate accuracy was 0.07 to 0.1 mag. from V = 13 to 17.   Forty to fifty stars (photoelectric and CCD photometry) up to V = 18 - 19 were used for the calibration.

3.2 Photographic Astrometry

3.2.1 Proper Motions

Proper motions were obtained from two plates scanned with the digitizing machine MAMA (Machine Automatique a Mesurer en Astronomie) at the Observatoire de Paris.   This machine was designed for astrometric purposes.   It is a fast machine; 1 cm^2 is scanned in 10 seconds, the repetivity is 0.1 micron (200 stars scanned 32 times), the rms error for a coordinate is less than 0.5 micron (Danguy, 1988).   The spot size is 10 microns, the step size used here is 10 microns (but can be reduced to 0.1 micron).   A 10 micron pixel size is required to detect and properly sample the dimmest stellar images.

3.2.2 Digital Centering

In searching for the optimal centering of stellar images for astrometric purposes, we have compared the efficiency of various algorithms (Bienaymé et al. 1988).   Five algorithms have been tested: 1) gravity center, 2) gradient method, 3) moment method, 4) 1D Gaussian fit to marginal density distribution and 5) maximum of the correlation function of the 2D stellar profile with the symmetric profile. Algorithms 1 through 4 are commonly used and the 1D Gaussian fit is recognized by many authors to be very efficient (Stone 1989) and to give centers nearly as good as a 2D Gaussian fit (Auer and van Altena 1978).   The fifth algorithm is a two dimensional one and the centroid is defined by coordinates a and b where F(a,b) is maximum and

$$F(a,b) = \iint f(x,y) \; f(a - x, \; y - b) \; dxdy$$

where f(x,y) is the density profile of the stellar image.

These algorithms have been tested comparatively on synthesized images for which the photon statistics are known.   Then we have compared, under various conditions, the efficiency of each method to the Cramer-Rao limit that is the limiting centering accuracy deduced from statistical considerations.

We conclude that as long as the signal to noise ratio remains high each algorithm gives comparable results in agreement with the Cramer-Rao limit, while at low S/N only algorithms number 4 and 5 give satisfactory results.   The 1D Gaussian fit could be retained since it is fast.   However we found that, for real CCD images or Schmidt plates, the correlation algorithm is more robust when a defect or a neighboring star is present inside the analyzed window.   The reason is that the computed center is in fact the center of the symmetric part of the image and that an unsymetric component, such as a defect, contributes little in the computation of F(a,b).

3.2.3 Null Proper Motions.

Scanning and centering on two similar Schmidt plates (same night, color, field...) have been done. Since the proper motions of stars in this case are zero, the measured proper motions reflect the errors due to scanning and centering. We found a dispersion of 1.5 microns (or 0.1") in proper motion corresponding to a 1 micron centering error on each plate and each coordinate. The variation in centroid determination by different algorithms working on the same stellar image is in fact smaller. Then we reasonably assume that the dispersion observed in null proper motion is mainly related to the noise statistics of the stellar images and could not be greatly reduced.

3.2.4 Plate to Plate Transformation.

The proper motion solution is strictly differential. Strictly speaking, there is no plate solution nor did we introduce the so-called standard coordinates. When the two catalogues of position (x,y) 1955 and (X,Y) 1986 for stars on each plate were built, the transformations X(x,y) and Y(x,y) between the two coordinate systems and proper motions were:

$$\mu_x = X_{86} - X_{55}$$
$$\mu_y = Y_{86} - Y_{55}$$

A simple transformation with only linear terms, such as X = a + bx + cy, appears to be insufficient to model the distortions between the two catalogues. In that case the mean proper motion in various areas of the catalogue varies between ±60 microns over the field, meaning that distortions (due to the fact that plates come from different telescopes, that in one plate the field is near the center and in the other near the edge, that one plate is a glass copy) are incorrectly modeled. To improve the method we write X and Y as being Chebichev polynomial expansions.

$$X(x,y) = \Sigma \ a_{i,j} \ T_i(x) \ T_j(y)$$

where T_i is a Chebichev polynomial of order i. The order of X is N and the number of coefficients is $(N + 1)^2$. This procedure is strictly equivalent to the P. Brosche et al. (1989) one using Legendre or Hermite polynomials.

The $a_{i,j}$ coefficients were determined by a least squares procedure assuming that the mean proper motion of the catalogue was zero. So we measured only relative proper motions and assumed that the absolute mean proper motion is a constant. In particular, we expected that there is no star streaming in any part of the field.

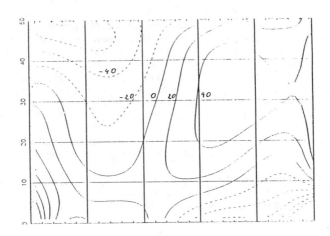

Fig. 1. Non-linear plate to plate x-distortions modeled with a fifth order transformation. There is a 20 micron step between two lines.

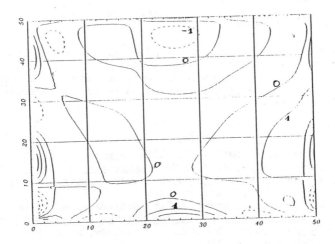

Fig. 2. Added x-distortions modeled by a sixth transformation relative to the fifth order. There is a 1 micron step between two lines.

For a given order N of the transformation, we checked that the mean proper motion of stars remained zero even in a small area of 1 cm^2 (i.e. 0.03 square degree) and that the dispersion of residuals (i.e., of proper motions) was constant. We could increase the order of the transformation to N + 1 and see if the residuals decreased significantly. Increasing the order of the transformation to N + 1 allows one to model distortions with a smaller scale length. This could be done only if the number of stars in the catalogue is sufficient. Fig. 1 shows non-linear x-coordinate distortions modeled between the two plates for N = 5. Fig. 2 shows added x-corrections between order N = 6 and N = 5; the extreme corrections are 1 micron.

Finally we found that order 4 or 5 gave the most significant coefficients. We tried up to order 9 (100 coefficients) and the proper motions remained quite unchanged. The accuracy in proper motions was deduced to be 0.2"/century.

4. CONCLUSION

A large proper motion survey could be done using Schmidt plates and a 0.2"/century accuracy would be reached with a time baseline of 31 years. A POSS plate copy has been used successfully as a first epoch and plate distortions can be modeled and corrected to compute the proper motions. Differential astrometry can be done between plates from different instruments (if a sufficient number of stars with good centering are measured on each plate). The best centering was obtained if smallest stellar images were digitized with a 10 micron step size. The best algorithm was not necessarily the 1D Gaussian fit. The 2D method can be more efficient.

A Preliminary analysis of the 1.8 square degree survey near M 5 shows qualitative evidence of a thick disk. A quantitative measurement of thick disk properties could be obtained with a 20 square degree survey. This is a straightforward extension of the reduction described above and is under way.

REFERENCES

Auer, L. H. and Van Altena, W. F. 1978 Astron. J. 83, 531.
Bahcall, J. and Soneira, R. M. 1984 Astrophys. J. Suppl. 55, 67
Bienaymé, O., Robin, A. and Crézé, M. 1987 Astron. Astrophys. 180, 94.
Bienaymé, O., Motch, C., Crézé, M. and Considère, S. 1988 in IAU Symposium No. 133, Mapping the Sky - Past Heritage and Future Directions, S. Debarbat, J. A. Eddy, H. K. Eichhorn, eds., Kluwer, Dordrecht, p. 389.
Brosche, P., Wildermann, E. and Geffert, M. 1989 Astron. Astrophys. 211, 239
Danguy, T. 1988 Note Technique INSU.
Murray, C. A., Argyle, K. W. and Carbon, D. M. 1986 Monthly Not.

Roy. Astron. Soc. 223, 629.
Ratnatunga, K. U., Bahcall, J. N. and Casertano, S.   1989   Astrophys.
    J. 339, 106.
Robin, A. and Crézé, M.   1986   Astron. Astrophys. 157, 71.
Robin, A., Crézé, M. and Mohan, V.   1989   in Evolutionnary Phenomena
    in Galaxies, J. Beckman. ed., Kluwer Academic Publishers,
    Dordrecht, in press.
Robin, A., Crézé, M., Bienaymé, O. and Oblak, E.   1989
    in The Force Perpendicular to the Galactic Plane, A. G. D.
    Philip and P. K. Lu, eds., L. Davis Press, Schenectady, p. 33.
Robin, A. and Oblak, E.   1987   in Proc Xth IAU European Astronomy
    Meeting, Vol 4, J. Palous, ed., p. 323.
Stone, R. C.   1989   Astron. J. 97, 1227.

THE MASS DISTRIBUTION IN THE GALACTIC DISK

Gerard Gilmore and Konrad Kuijken

Institute of Astronomy, Cambridge

ABSTRACT: The   recent   determination   of   the   integral   surface   mass
density   of   the   Galactic   disk,   together   with   reanalysis   of   the
uncertainties   in   available   determinations   of   the   local   volume   mass
density   near   the   Sun,   show   that   there   is   no   evidence   for   any
unidentified mass associated with the Galactic disk.   Correspondingly,
about one-half of the total mass interior to the Sun must be in an
extended halo of unknown material.

1.   INTRODUCTION

The distribution of mass in the Galactic disk is characterized by
two numbers, its local *volume* density ρ_0 and its total *surface* density
$\Sigma(\infty)$.   Both these dynamical quantities are derived from a measurement
of the vertical Galactic force field, $K_z(z)$.   Although $\Sigma(\infty)$ and ρ_0 are
different measures of the distribution of mass in the Galactic disk
near the Sun, they are related.   Of the two, the most widely used and
commonly determined measure is the local *volume* mass density - i.e. the
amount of mass per unit volume near the Sun, which for practical
purposes is the same as the volume mass density at the Galactic plane.
This quantity has units of $M_{\odot}$ pc^{-3}, and its local value is often called
the "Oort limit" in honor of the early attempt at its measurement by
Oort   (1932).   The   contribution   of   *identified*   material   to   the
dynamically determined Oort limit may be determined by summing all
local observed matter - an observationally difficult task, which leads
to considerable uncertainties.   The uncertainties arise in part due to
difficulties in detecting very low luminosity stars, even very near the
Sun, in part from uncertainties in the binary fraction among low mass
stars, in part from uncertainties in the stellar mass - luminosity
relation, but mostly from uncertainties in determining the volume
density of the interstellar medium (ISM).   This latter uncertainty is
exacerbated since the physically important quantity for dynamical
purposes is the mean volume density of the patchily distributed ISM at
the solar galactocentric distance.   The best available determination of
the local mass density in identified material is ~ 0.1 $M_{\odot}$ pc^{-3}, with a
very-poorly defined uncertainty of perhaps as much as 25% in this

A. G. Davis Philip and P. K. Lu (eds.)
The Gravitational Force Perpendicular to
the Galactic Plane   61 - 80
©   1989   L. Davis Press

value.   Comparison of this value with that determined from dynamical analyses is required to test for the existence of dark matter associated with the Galactic disk.

The second measure of the distribution of mass in the solar vicinity is the integral surface mass density. This quantity has units of $M_\odot$ pc^{-2}, and is the total amount of disk mass in a column perpendicular to the Galactic plane.   It is this quantity which is required for the interpretation of rotation curves and the large-scale distribution of mass in galaxies.   Recent determinations of this surface mass density lead to values in the range 45 $M_\odot$ pc^{-2} to 80 $M_\odot$ pc^{-2}.   As an indication of the global dynamical significance of this mass density, the contribution of a disk potential generated by some known local mass density to the local circular velocity, assuming an exponential disk with the Sun 2.5 radial scale lengths from the Galactic center, is

$$V_{c,disk} \sim 150 \left(\frac{\Sigma_{local}}{60 \mathcal{M}_\odot pc^{-2}} \right)^{\frac{1}{2}} km\,s^{-1}. \qquad\qquad 1$$

The local circular velocity is ~220 km/s. The contribution of the potential due to a given enclosed mass M to the circular velocity is approximately $V_c^2 = GM/r$, so that contributions to the observed local circular velocity from the various mass components generating the Galactic potential add in quadrature.   Thus the Galactic disk provides only about 50% of the total Galactic potential at the solar galactocentric distance.   The nature of the mass which generates the other half of the potential remains unknown.

If one knew both the local *volume* mass density and the integral *surface* mass density of the Galactic disk, one could immediately constrain the scale height of any contribution to the local volume mass density which was not identified.   For example, one might suspect that some fraction of the local volume mass density was unidentified (i.e. a local "dark mass" problem), but also determine a surface density which is effectively fully explained by observed mass.   Then the unidentified contribution to the local volume density would have to have a small scale height, in order that its integral contribution to the surface density be small.   In view of the very small scale height on which it must be distributed, it would then be plausible to deduce that any "local" dark mass unidentified in the volume mass density near the Sun was not the same "dark" mass which dominates the extended outer parts of galaxies.

2. MEASUREMENT OF THE GALACTIC POTENTIAL

Determination of the volume mass density and the integral surface mass density near the Sun require similar observational data, namely distances and velocities for a suitable sample of tracer stars, but rather different analyses.

All determinations of the mass distribution in the Galactic disk require a solution of the collisionless Boltzmann equation. In view of the inconvenience of general solutions of this equation derived from real data, in practice one utilizes its vertical velocity moment, the vertical Jeans equation:

$$K_z = \frac{1}{\nu}\frac{\partial}{\partial z}(\nu \sigma_{zz}^2) + \frac{1}{R\nu}\frac{\partial}{\partial R}(R\nu\sigma_{Rz}^2)$$ 2

where $\nu(R,z)$ is the space density of the stars, and

$$\vec{\sigma}_{ij}(R,z) = \langle v_i v_j \rangle - \langle v_i \rangle \langle v_j \rangle$$

their velocity dispersion tensor.

The first term on the right hand side of equation 2 is dominant, and contains a logarithmic derivative of the stellar space density $\nu(z)$, and a derivative of the vertical velocity dispersion, σ_{zz}. Since the stellar population in the solar neighborhood is, within a multiplicative factor of a few, tolerably well described by an isothermal stellar population, the part of this term containing the derivative of the space density dominates the determination of $K_z(z)$ near the Sun. This point is not often appreciated adequately, but means that one should determine stellar density profiles with even greater care than that required for the velocity dispersions.

The second term in the Jeans equation 2 describes the tilt of the stellar velocity ellipsoid away from the local cylindrical-polar coordinate system in which velocity dispersions are measured. One therefore needs the R-gradients of σ_{Rz} and of ν. There are no general analytical solutions for this term, as it depends on the unknown "third integral" of the motion. Estimates of its importance may be derived by numerical integration of orbits in potentials which are thought to be realistic approximations to that of the Galaxy. To illustrate the significance of this non-separability in determining the σ_{Rz}-term, Fig. 1 shows the projections in the (R,z)-plane of three orbits computed in the disk-halo potential described by Carlberg and Innanen (1987). The envelopes of these orbits agree in the plane, but diverge slightly at large z. Consequently different orbits fill out boxes which are not part of a single coordinate grid, and this potential is not separable. Note that these orbits tilt towards the Galactic center, but not quite as strongly as they would have in a spherical potential, when they would have been aligned with the dashed lines in Fig. 1. Thus one may use numerical orbit integrations to estimate the likely range of values for the σ_{Rz}-term in the Jeans equation, while Fig. 1 shows that the value of this term in a spherical potential will be quite close to that in the real Galaxy. Quantifying the precision of this last statement is not a trivial exercise.

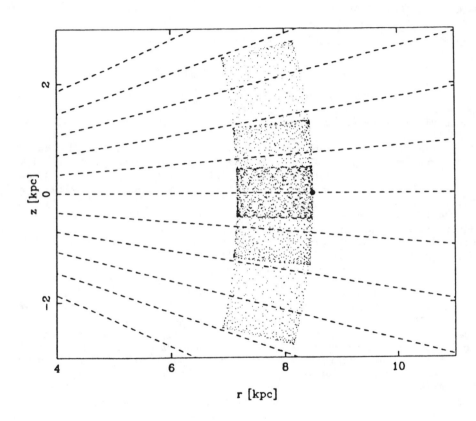

Fig. 1.   The projection in the (R,z)-plane of three orbits calculated
in a plausible Galactic potential.   The dashed lines pass through the
Galactic center.   The envelopes of these orbits are well aligned with
the Galactic radius vector in the plane, but diverge slightly from it
at large z, providing an estimate of the orientation of the stellar
velocity ellipsoid far from the plane.

One may however derive a realistic upper limit on the importance of the second term in the Jeans equation by considering velocity ellipsoids which are oriented towards the Galactic center. In this case, if the disk of the Galaxy is self-gravitating, radially exponential and has a constant vertical scale height, as is seen in external disk galaxies, vertical balance implies (for disk surface density μ) $\sigma_{zz}^2 \propto \mu$, and hence

$$\sigma_{zz}^2 \propto \mu \propto \nu \propto e^{-R/h_R}. \qquad\qquad 3$$

Thus we obtain

$$\frac{1}{R\nu}\frac{\partial}{\partial R}(R\nu\sigma_{Rz}^2) = 2(\alpha^2 - 1)\sigma_{zz}^2 \left(\frac{\alpha^2 z^3}{(\alpha^2 z^2 + R^2)^2} - \frac{Rz}{h_R(\alpha^2 z^2 + R^2)}\right) \qquad 4$$

as the tilting term for a radially exponential population of constant vertical scale height with a velocity ellipsoid of constant axis ratio $\sigma_{RR} : \sigma_{zz} = \alpha$ which points at the Galactic center (Kuijken and Gilmore 1989a). Since this term is proportional to σ_{zz}, inserting it into the Jeans equation 2 gives a linear equation in σ_{zz}, from which one can deduce K_z. The tilt term is clearly unimportant near the plane, as $z \Rightarrow 0$, and may legitimately be ignored in analyses restricted to local determinations of ρ_0.

3. FROM GRAVITY TO MASS

Given a measurement of the gravitational field $\vec{K}(R,z)$ in an axisymmetric galaxy, the total density ρ of gravitating matter follows from Poisson's equation:

$$\nabla \cdot \vec{K} = -4\pi G\rho. \qquad\qquad 5$$

In the case of a disk galaxy we can express the R-gradient in $\nabla\cdot\vec{K}$ in terms of the observed circular velocity at the Sun, v_c or in terms of the Oort constants of Galactic rotation A and B

$$\rho = -\frac{1}{4\pi G}\left\{\frac{\partial K_z}{\partial z} + \frac{1}{R}\frac{\partial}{\partial R}(RK_R)\right\}$$

$$= -\frac{1}{4\pi G}\left\{\frac{\partial K_z}{\partial z} + \frac{1}{R}\frac{\partial (v_c^2)}{\partial r}\right\}$$

$$= -\frac{1}{4\pi G}\left\{\frac{\partial K_z}{\partial z} + 2(A^2 - B^2)\right\}. \qquad 6$$

For a disk galaxy with an approximately flat rotation curve the second term is small within a few kpc of the disk plane (Kuijken and Gilmore 1989a provide a more exact calculation; for an exactly flat rotation curve $A^2 - B^2 \equiv 0$ at $z = 0$), so we can integrate in z to obtain the total column density $\Sigma(z)$ between heights $-z$ and z relative to the disk plane $z = 0$:

$$\Sigma(z) = \int_{-|z|}^{|z|} \rho(z)dz = \frac{|\mathcal{K}_z|}{2\pi G} - \frac{(A^2 - B^2)}{\pi G}|z|. \qquad\qquad 7$$

The physical interpretation of a determination of the Galactic $K_z(z)$ force law can now be seen from inspection of equations 2, 6 and 7. In effect, one measures the pressure - gravity balance of the *collisionless* stellar "fluid". The hydrodynamic analogy following from the description of the collisionless Boltzmann equation as the equation of stellar hydrodynamics is particularly appropriate here. The dominant first term on the right hand side of equation 2 contains a logarithmic derivative of the stellar spatial density $\nu(z)$, and the stellar velocity dispersion σ_{zz}. The spatial density term plays the role of a scale height, the velocity dispersion is analogous to a temperature, and the product $\nu\sigma_{zz}$ is a pressure.

Some complexity is introduced since the stellar fluid is collisionless. This means that non-diagonal terms in the velocity dispersion tensor exist [the second term on the right hand side of equation 2] in consequence of the fact that pressure is not isotropic, and we have no equation of state to close the series of moment equations. It also means that one must measure the temperature-density balance locally by point-by-point sampling of many stars, imposing some inconvenience on the observational techniques, but conversely allowing the possibility of more sophisticated analyses.

It is evident from the equations above that determinations of the local volume mass density ρ_0 depend on the square of any distance scale errors in the tracer population, since they are derived from the second derivative of the stellar space density distribution, while determinations of the surface mass density are linearly proportional to the distance scale, being based on the first derivative. Since the stellar space density is itself a derivative from the basic star count data, it is evident that determinations of $K_z(z)$ and particularly of ρ_0 are very sensitive to sampling noise (and systematic errors) in the data. Direct analyzes of combined density law and velocity dispersion data which proceed by substituting observed values into equations 2 and 6 are therefore prone to produce rather erratic answers.

4. ISOTHERMAL DECONVOLUTIONS

Almost all determinations of ρ_0 model the distribution function of the tracer stars as one or more *isothermal* components. An isothermal component by definition has constant velocity dispersion, independent of z. Then one can easily see that the velocity distribution has to be Gaussian: by equation 2, if σ_{zz} is independent of z,

$$\nu(z) \propto e^{-\psi(z)/\sigma_{zz}^2},$$ 8

and hence the Abel inversion described below (equation 17) yields

$$f_z(E_z) \propto e^{-E_z/\sigma_{zz}^2} \propto e^{-\frac{1}{2}v_z^2/\sigma_{zz}^2}.$$ 9

This method is especially simple if the tracer appears to be a single isothermal component and the analysis is restricted to stars near the plane, as then the Jeans equation simplifies to

$$\mathcal{K}_z = \sigma_{zz}^2 \frac{\partial}{\partial z}\left(\ln \nu(z)\right)$$ 10

and only the density gradient of the tracer population is needed.

The reliable deconvolution of non-isothermal populations into isothermal components is neither trivial nor robust. First, there is the question of how physically meaningful it really is to separate a stellar sample into discrete isothermal subsamples. Given continual star formation and the continuous diffusion processes thought to be active in heating the disk, this can at best be only an *ad hoc* approximation to the true distribution. A very large number of discrete isothermals is clearly necessary for a close approximation to a real galactic disk. Second, such deconvolutions are far from unique, leading to non-unique force laws. There is, however, no difficulty *in principle* with such deconvolutions; the difficulties lie entirely in the very large amounts of data necessary to make the deconvolution reliable.

At distance z_j, a superposition of N_{iso} isothermal components, each with velocity dispersion σ_{zz}^2 and density ν_i at $z = 0$ has density

$$\nu(z_j) = \sum_{i=1}^{N_{iso}} \nu_i e^{-\psi(z_j)/\sigma_{zz,i}^2}.$$ 11

This makes clear that at different distances from the plane the relative densities of the components are not the same, but that those of highest velocity dispersion dominate increasingly at higher z. If we were able to specify the velocity dispersions and the spatial density normalization at $z = 0$, $\nu_i(0)$ of all components, this would allow a direct solution for the potential from the density profile. However in practice we have to derive the number density of the high-velocity dispersion components at $z = 0$ from high-z velocity data, as only at greater heights are these components sufficiently in evidence to allow reliable determination of their density. However, for $z \neq 0$, the relative densities of the isothermal components depend on the potential as well as the ν_i; thus we are faced with a self-consistency problem involving both the density profile and the potential far from the plane even if we wish to solve for the potential only near the plane.

At height z_j, the mth velocity moment of a superposition of isothermals is

$$\langle |v_z|^m \rangle (z_j) = \frac{\sum_i \nu_i e^{-\psi(z_j)/\sigma_{zz,i}^2} \left(\frac{2\sigma_{zz,i}^2}{\pi}\right)^{m/2} \left(\frac{m-1}{2}\right)!}{\sum_i \nu_i e^{-\psi(z_j)/\sigma_{zz,i}^2}}. \qquad 12$$

We could thus proceed by specifying a set of N_{iso} velocity dispersions $\{\sigma_{zz,i}^2\}$, and fit the densities and velocity dispersions observed at each of N_z heights z_j, solving for the potential and for the density normalizations of the individual components. However, without parameterizing the potential in some way, there are many (highly correlated) degrees of freedom in such fits: equations 11 and 12 give $2N_z$ constraints for $(N_z + N_{iso} - 1)$ fit parameters (the -1 arises from the free zero point of the potential). Typically, in past applications of this method, $N_z \approx 10$ and N_{iso} would have to be at least three to be able to describe the data adequately. With a small number of components, it turns out (not surprisingly) that the derived potential is quite sensitive to the dispersions of the individual model components. Since we require the second derivative of this potential considerable uncertainty is inevitable.

5. SELF CONSISTENT SOLUTIONS FOR ρ_0

An almost invariable finding in early studies was that a *maximum* was found in the $K_z(z)$ law a few hundred pc from the plane. Such a result is physically impossible, corresponding to a layer of negative mass. The solution to this inconvenient situation was found by Oort (1960), who imposed consistency with the Poisson equation 2 on his solution of equation 10. In effect, one assumes that uncertainties in the data rather than in the assumptions underlying the analysis are at fault, and that the theoretically reasonable solution which is most consistent with the (defective) data will produce an answer which is close to the "true" answer. Assignment of a realistic uncertainty to the resulting answer is of course somewhat problematic in this case. In Oort's (1960) reanalysis of Hill's (1960) K-giant data, he calculated the fractions of the various components using a first-pass guess at the potential, rather than solving for them and the potential fully self consistently. Oort used three isothermal components in his fit to the K-giant data, so that a consistent solution was possible.

More recently, Bahcall (1984a,b,c) has extended this idea, and has developed a substantial improvement in the theoretical methods with which to determine the local volume density of matter ρ_0, by deriving a joint solution of the Poisson and collisionless Boltzmann equations. This solution requires the derived $K_z(z)$ law to be physically possible, which is a highly desirable and quite reasonable constraint. The analytical techniques developed by Bahcall (1984a,b,c) represent a considerable improvement over those applied previously, and for the first time allow a derivation of ρ_0 which is limited by the quality of the available observational data, rather than by the approximate nature of the analysis.

In this technique, the solar neighborhood is divided into different isothermal components, each of which responds to the potential ψ via its velocity dispersion:

$$\rho_i = \rho_{i,0}\, e^{-\psi/\sigma^2_{zz,i}}.$$ 13

Self-consistency of the potential and the total matter density in these components then requires that Poisson's equation be satisfied, i.e.

$$4\pi G\rho = 4\pi G\sum_i \rho_{i,0}\, e^{-\psi/\sigma^2_{zz,i}} = \frac{d^2\psi}{dz^2}.$$ 14

Here the density includes a constant "effective halo" density, due to the halo mass and the radial gradients of the global gravitational field of the Galaxy - this term will be discussed further below. With the boundary conditions $\psi = \psi' = 0$ at $z = 0$, equation 14 can easily be integrated forwards to obtain $\psi(z)$. A variety of dark matter components can be added in, using the same prescription as for the visible components, and the resulting potentials calculated.

A critical problem is to analyze the gas and stars into isothermal components; this can be done to reasonable precision very near the plane, but at higher z the calculated potential becomes increasingly sensitive to the precise $z = 0$ velocity dispersions and densities. The modeling of the gas is also a problem, as it accounts for about one-half of the locally identified volume density, but its precise density and spatial distribution are poorly known.

Bahcall (1984a,b) used his algorithm to reanalyze the available F-dwarf and K-giant high-Galactic latitude data with new models which are self-consistent in the sense that the matter which generates the gravitational field itself responds to it in a manner described by the collisionless Boltzmann equation. Bahcall found that:

(i) the gravitational field due to the 0.10 $M_\odot$ pc^{-3} of stars and gas that are identified in the solar neighborhood is inconsistent with the gravitational fields derived from the data;

(ii) depending on its scale height, a further 0.06 - 0.14 $M_\odot$ pc^{-3} of unidentified matter is required. This unidentified matter is not part of a very extended halo, though that must also exist and have a local volume mass density of ~ 0.01 $M_\odot$ pc^{-3} so that there is sufficient mass in the Galaxy to generate the potential required to explain the local circular velocity.   Hence this result implies significant amounts of disk-like, dissipational dark matter in the solar neighborhood.

6.   UNCERTAINTIES IN THE LOCAL VOLUME MASS DENSITY

    In view of the important consequences of the existence of large amounts of dissipational dark mass, it is of interest to examine the uncertainties in the determination of the Oort limit.   The sensitivity of determinations of the local volume mass density ρ_0 to uncertain data lies in the modeling of the stellar velocity distribution near the Galactic plane, and in the determination of the stellar-density distribution with distance from this plane.   Both F-dwarf and K-giant tracer samples have been analyzed to determine ρ_0, with both producing a result of ρ_0 ~ 0.20 $M_\odot$ pc^{-3}, where the identified mass provides $\rho_{0,obs}$ = 0.10 $M_\odot$ pc^{-3} (Bahcall 1984c).

    The effect of *random* errors on determinations of the Oort limit can be investigated using Monte Carlo simulations of the data acquisition and analysis.   Random errors can produce both random and systematic effects on a dynamically-determined quantity, through effects similar to Malmquist bias.   It is therefore important to understand the effects of such errors.   Analyses of this type have been undertaken, but unfortunately disagree in their conclusions.   Gilden and Bahcall (1985) conclude that random errors produce an unbiased uncertainty of ~ 1% in ρ_0, while Bienaymé, Robin and Créze (1987) and Crézé, Robin and Bienaymé (1989) conclude that random errors produce an uncertainty of ~ 50%, and also produce a bias towards an erroneous detection of unidentified mass.   The difference in these results is due to different techniques for handling observational errors in the simulations, suggesting that the appropriate uncertainty to apply to determinations of the Oort limit is not yet well quantified.

    An alternative, and perhaps more objective, method for the determination of an error bar is to do the experiment with two different but supposedly similar samples of stars, and compare the resulting answers.   Fortunately for this purpose, the F-star sample analyzed is the sum of two sub-samples (F 5 and F 8, Hill et al. 1979), with no evidence for a difference between their velocity distributions (Adamson et al. 1988).   For steady-state stellar populations, two tracer populations with the same kinematics in the same gravitational potential must follow the same spatial density distribution. For the F 5 and F 8 samples the data shown in Fig. 2 show that this is not the case.   One or both of the data or the assumptions underlying the modeling of the F-star kinematics is thus clearly in error.   The amplitude of the resulting uncertainty can be found by deducing ρ_0 from each of the three F-star samples, F 5, F 5 + F 8 and F 8, using the algorithm derived by Bahcall (1984a).   The resulting values of ρ_0 are

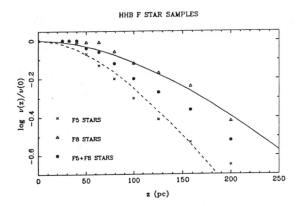

Fig. 2. The Hill et al. (1979) F-star samples. The difference between the density profiles of the F 5 and the F 8 samples is evident. The curves show separate model fits calculated using the algorithm devised by Bahcall (1984a) to the F 5 and the F 8 subsets of the data defined by Hill et al. Only the averaged sample (solid points) was analyzed by Bahcall (1984b). The models shown have local volume mass densities: ρ_0 = 0.11 $M_\odot$ pc^{-3}, i.e. with no missing mass (solid line), and ρ_0 = 0.29 $M_\odot$ pc^{-3} (dashed line).

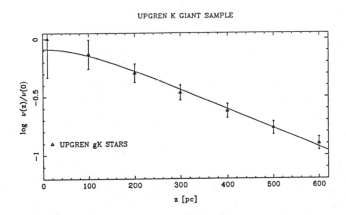

Fig. 3. Weighted fit to the Upgren (1962) K giant data, using the velocity distribution measured by Hill (1960). The model shown contains no dark matter in the Galactic disk, and has ρ_0 = 0.10 $M_\odot$ pc^{-3}.

0.29 $M_\odot$ pc^{-3}, 0.185 $M_\odot$ pc^{-3} (reproducing the result derived by Bahcall (1984b) exactly), and 0.11 $M_\odot$ pc^{-3} respectively. Thus one may deduce that there is twice as much mass missing as observed in the local volume density, just as much missing as observed, or no missing mass at all, depending on which sample of stars one chooses to analyze. Clearly, the available F-star data are not capable of providing any evidence either for or against the concept of missing mass near the Sun.

The sample of K-giants which has been analyzed previously has been shown to have a velocity distribution which is consistent with a single isothermal, with a velocity dispersion of ~ 20 km/s (Bahcall 1984c). Thus, unlike the F-stars, in this model the K-giants consist entirely of old disk stars, with neither young disk nor thick disk star representatives. Since stars of a wide range of masses become K-giants, including the present F dwarfs, this model is inherently implausible. Remember also that Oort (1960) found it necessary to model the same K-giant sample as the sum of three isothermal components, with (crudely) one component each for the young disk stars, the old disk stars, and the high velocity halo stars. A further complication follows from a feature of Bahcall's analysis, which assigns high weight to the density profile near the plane, where the number of stars counted is smallest. Reanalysis of published data including weighting of the density data by its Poisson noise and using the detailed fit to the local K-giant velocity data derived by Hill (1960), leads to a value of ρ_0 = 0.10 $M_\odot$ pc^{-3} (Fig. 3, Kuijken and Gilmore 1989c). The previously derived value from the same data using the same analysis technique was ρ_0 = 0.21 $M_\odot$ pc^{-3} (Bahcall 1984c).

The fundamental reason for the difficulty with analysing data by deconvolving the tracer sample into a few isothermal populations has been outlined by Fuchs and Wielen (1989). Basically, it is physically impossible for such a thing as a real "isothermal" stellar population to exist, since stars are formed and evolve dynamically as a continuum. Thus any "isothermal" stellar group must include a range of ages and kinematics. Determination of the local volume mass density is much more than linearly sensitive to deviations from isothermality. Hence undetectable deviations of a tracer sample from the idealized assumptions of the analysis produce significant errors in the resulting volume mass density. Fuchs and Wielen (1989) have calculated the amplitude of this effect. They show that the known rate of kinematic diffusion for stars near the Sun operating for the age range corresponding to the observed F-star sample means that the apparently isothermal F-star sample, which has an observed velocity dispersion of ~ 11.5 km/s, can be better modeled as having a range of velocity dispersions from 10 km/s to 13 km/s. This apparently small change in the model results in a change in the derived local volume mass density from 0.19 $M_\odot$ pc^{-3} to 0.10 $M_\odot$ pc^{-3}. Thus the results of such an analysis are extremely sensitive to the precision of the adopted kinematic model.

7. DETERMINATION OF THE SURFACE MASS DENSITY

Because high-energy stars are present at all heights above the Galactic plane, measurements of the potential very close to the plane still require knowledge of the high-energy tail of the distribution function. Therefore either the tail of the velocity distribution at low z, or the density *and* potential at high z, are required to measure the potential at low z, and hence to deduce the local volume density of matter ρ_0. The phase-space distribution function we discuss below, however, depends on the density only at points farther from the plane than the height at which data are being analyzed. It is possible to capitalize on this insensitivity to the detailed shape of the potential (equivalently, the detailed mass distribution) near the plane to derive the potential at large distances from the plane from high-z data alone. Since a measurement of K_z at any height relates directly to the total surface density integrated to that height, this is extremely useful, allowing us to obtain meaningful determinations of the surface mass density of the Galactic disk from high-z data alone.

In most K_z-studies, the density $\nu(z)$ is known to better precision than the velocity distribution. Instead of fixing the parameters of the latter, and then using these to model the density gradient, it is therefore preferable to work in the other direction, and predict the velocity distribution of a tracer in different model potentials, given its density. These velocity distribution models can then be compared with the observed velocity data using maximum likelihood techniques.

Given a distribution function $f_z(E_z)$ and a potential $\psi(z)$, we can calculate the density $\nu(z)$, which is just a moment of f_z:

$$\nu(z) = \int_{-\infty}^{\infty} f_z(z, v_z) dv_z$$

$$= 2 \int_{\psi(z)}^{\infty} \frac{f_z(E_z)}{\sqrt{2(E_z - \psi(z))}} dE_z. \qquad 15$$

Reparameterizing the z-height in terms of the potential ψ, we have

$$\nu(\psi) = 2 \int_{\psi}^{\infty} \frac{f_z(E_z)}{\sqrt{2(E_z - \psi)}} dE_z. \qquad 16$$

This equation is an Abel transform, which has the well-known inversion (see Binney and Tremaine 1987):

$$f_z(E_z) = \frac{1}{\pi} \int_{E_z}^{\infty} \frac{-d\nu/d\psi}{\sqrt{2(\psi - E_z)}} d\psi, \qquad 17$$

so that there is a unique relation between $\nu(\psi)$ and $f_z(E_z)$. Because of this equivalence of $\nu(\psi)$ and $f_z(E_z)$, there is a triangular mathematical relationship between the three functions $\psi(z)$, $\nu(z)$ and $f_z(E_z)$: any one of them can be deduced from the other two. Abel inversions are somewhat unstable, but not as unstable as taking a direct derivative of the data.

It is important to note that equation 17 shows that $f_z(E_z)$ depends on the density only at points where the potential exceeds E_z, i.e. beyond the point $z = \psi^{-1}(E_z)$. It is this property which allows the derivation of $K_z(z)$ at large z independently of the poorly known distribution of mass near the plane.

When starting from a set of (z, v_z) data for the tracer population, only the first of the three quantities $\nu(z)$, $f_z(E_z)$ and $\psi(z)$ is known, as one needs the potential to be able to convert $f_z(z,v_z)$ into $f_z(E_z)$. Therefore, before being able to make an inversion such as that given in equation 17 we have to make some assumption about the form of the potential, or about $f_z(E_z)$. Assuming that the tracer is isothermal ($f_z \propto e^{-E_z/\sigma_{zz}^2}$) is an example of the latter, leading to a trivial inversion to equation 11.

An analysis technique based on equation 17 has been devised by Kuijken and Gilmore (1989a,b) for the determination of $K_z(z)$, and $\Sigma(z)$. The essential feature of that analysis is that one avoids the assumption of isothermality, by instead postulating a range of potentials $\psi(z)$, and for each of them calculating $f_z(z, v_z)$ from $\nu(z)$. The range of model distribution functions can then be compared to the observed distribution function of velocity - distance data, and used to select the best-fitting model potential.

This is not a direct measurement of the potential, but rather a modeling of it, so it is important to make sure that the models are sufficiently general. On the other hand, constraints can be built into the model potentials which direct derivations from data have more difficulty coping with. For example, the analysis by Hill (1960) of a K-giant sample found a $|K_z(z)|$ which started to decrease above a few hundred parsecs, which leads one to deduce negative dynamical masses - a physical absurdity. The same data were shown by Oort (1960) to be also consistent with a K_z-law which did not show such a turnover. This illustrates the fact that some force laws can be ruled out on physical grounds, and so should not be included among the possible solutions (provided, of course, that acceptable fits to the data can be obtained from the restricted solution set). Consistency with the Galactic rotation curve can also be built into the potentials: if one were to find a very heavy disk mass, for example, which can generate most of the local circular speed by its gravity, one would not expect also to find evidence for a massive halo in K_z.

8.   A SIMPLE PARAMETERIZATION OF PLAUSIBLE K_z FUNCTIONS

    As demonstrated above, $K_z(z)$ is related directly to the surface density to height z, $\Sigma(z)$. This means that the detailed distribution of the matter in the disk (whether it has a high or low scale height, whether its distribution is close to exponential or more like a sech^2-law,....) does not affect the high-z potential.   Hence one is able to derive the disk surface density without needing to know how the mass inside it is distributed, and can investigate simpler model potentials than those required for measurements of ρ_o.

    The total mass density along a slice perpendicular to the disk for a generic disk-halo system is shown schematically in Fig. 4.   (By "halo" in this context we mean any mass component which is not distributed like the dominant mass in the old disk. Nothing is implied as regards the nature of this mass; it includes the luminous "spheroid" or "bulge" as well as any roughly spherical distribution of dark matter.)   At low z, the density is mostly due to the disk component, while further away from the plane the density of the halo, which at large galactocentric radii is essentially constant over a few disk scale heights, dominates.   The surface density, or equivalently $|K_z|$, for such a system is also shown in Fig. 4.   This force law has the following generic features:

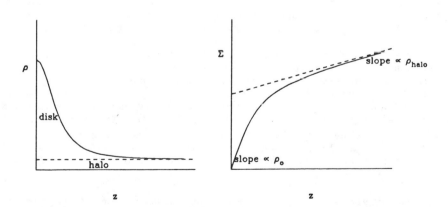

Fig. 4.   The volume density (a, left) and surface mass density, or equivalently K_z, (b, right) as a function of vertical distance z from the disk plane in a simple disk-halo model.

(i)   At low z, smaller than the scale heights of the dominant disk components, $|K_z|$ rises almost linearly, with slope $4\pi G\rho_o$.

(ii) At large z, beyond most of the mass of the disk, $|K_z|$ is again linear, but with a much reduced slope, equal to $4\pi G\rho_{halo}$.

(iii) The extrapolation of this latter linear portion back to $z = 0$ has an intercept of $2\pi G\Sigma_{disk}$.

An accurately measured K_z-profile over a wide range in z yields the disk surface density, as well as the volume density of the halo and that of the disk at $z = 0$.

The essential features of the behavior of the K_z-function may be quantified in terms of a disk scale height D, a disk surface density K, and a halo of local density F, as follows:

$$-K_z = 2\pi G\ K\ \frac{z}{\sqrt{z^2 + D^2}} + 4\pi G\ Fz, \qquad\qquad 18$$

corresponding to the potential

$$\psi(z) = 2\pi G\ K\ \left(\sqrt{z^2 + D^2} - D\right) + 2\pi G\ Fz^2. \qquad\qquad 19$$

These functions reproduce all the generic behavior of a disk-halo system, except for the details of the disk mass distribution near the plane, which are unimportant for determinations of $K_z(z)$ at large z.

The parameter F deserves further comment. It measures the large-z quadratic term in the potential, which is predominantly caused by the halo density (taken to be constant over a few disk scale heights). A part of it is due to the $(A^2 - B^2)$ term, however, discussed above. Because F behaves just like a constant halo density in the z-restriction of Poisson's equation, it has been termed the "effective halo density", ρ_{eff}. Although this name suggests that it is a function of the halo component only, this is misleading, because the quadratic part of the disk potential also contributes to the difference $\rho_{halo} - \rho_{eff}$.

Potentials of the form given by equation 19 can be combined with a measured density distribution $\nu(z)$ and equation 17 to predict a set of distribution functions of velocities as a function of distance. With appropriate care in handling observational errors, these can be compared with distance - velocity data using maximum likelihood to determine the most likely parameters in equation 19 to represent the Galactic potential. One may then derive the integral surface mass density of the Galactic disk. This technique has been applied by Kuijken and Gilmore (1989a,b), to determine a dynamical total surface mass density Σ_∞ = 46 $\pm$9 $M_\odot$ pc^{-2}. These same authors integrate the local observed volume mass density in stars through their derived potential, and add the directly observed mass in the interstellar medium, to derive an observed integral surface mass density of $\Sigma_{\infty\,obs}$ = 48 $\pm$ 8 $M_\odot$ pc^{-3}. There is thus evidence that there is no significant unidentified mass associated with the Galactic disk.

Determination of the appropriate uncertainties to apply to analyses of this type is extremely complex, as many of the parameters are correlated in subtle ways. Similarly, the amplitude of possible systematic errors, due for example to our limited knowledge of the gradient in the local rotation curve and of the shape and orientation of the stellar velocity ellipsoid at large distances from the Galactic plane, is extremely difficult to quantify. As with the determination of ρ_0, the most reliable determination of the systematic uncertainty will come from the comparison of the results described here with the results of a similar measurement based on an independent sample of stars, when such is available. Neither the random nor the systematic uncertainties in either the amount of identified mass or in the amount of dynamical mass are well determined, though both types of uncertainty are considerable. It is also true that absence of evidence is not evidence of absence. Occam's razor must be invoked, so that the minimal conclusion required by available data and analyses is clearly that there is no robust evidence for the existence of any dark mass associated with the Galactic disk.

REFERENCES

Adamson, A. J., Hill, G., Fisher, W., Hilditch, R. W. and Sinclair, C. D. 1988 Monthly Not. Roy. Astron. Soc. 230, 273.
Bahcall, J. N. 1984a Astrophys. J. 276, 156.
Bahcall, J. N. 1984b Astrophys. J. 276, 169.
Bahcall, J. N. 1984c Astrophys. J. 287, 926.
Binney, J. and Tremaine, S. 1987 Galactic Dynamics, Princeton Univ. Press, Princeton.
Carlberg, R. G. and Innanen, K. A. 1987 Astron. J. 94, 666.
Crézé, M., Robin, A. and Bienaymé, O. 1989 Astron. Astrophys. 211, 1.
Fuchs, B. and Wielen, R. 1989, in preparation.
Gilden, D. L. and Bahcall, J. N. 1985 Astrophys. J. 296, 240.
Hill, E. R. 1960 Bull. Astron. Inst. Netherlands 15, 1.
Hill, G., Hilditch, R. W. and Barnes, J. V. 1979 Monthly Not. Roy. Astron. Soc. 186, 813.
Kuijken, K. and Gilmore, G. 1989a Monthly Not. Roy. Astron. Soc. 239.
Kuijken, K. and Gilmore, G. 1989b Monthly Not. Roy. Astron. Soc. 239.
Kuijken, K. and Gilmore, G. 1989c Monthly Not. Roy. Astron. Soc. 239.
Oort, J. H. 1932 Bull. Astron. Inst. Netherlands 6, 249.

Oort, J. H. 1960 Bull. Astron. Inst. Netherlands 15, 45.
Upgren, A. R. 1962 Astron. J. 67, 37.

DISCUSSION

King: It bothers me that you derive your density law at an early stage and then take it as fixed forever.

GILMORE: The density law is not really fixed forever. It is unavoidable that something is fixed, as there is a triangular relationship between density, velocity and potential. One must fix one of these to derive the second by fitting the third. We therefore do the analysis with our basic best estimate of the density law, and then repeat the whole thing with a different density law which we estimated to include the likely maximum systematic change possible in our distance scales.

RATNATUNGA: What exactly do you use to remove K-giants contaminating the sample of k-dwarfs?

GILMORE: Our spectra include the gravity sensitive MgH feature, and all the stars are sufficiently cool that dwarf-giant discrimination is effectively error-free. In spite of that, we exclude all the bright stars from our analysis to minimize any possible problem, since all the giants are bright to a very good approximation.

GOULD: Is all the information in the K_z-law graph in the slope and the intercept of the asymptotic part?

GILMORE: The K_z law derived here is constrained by our data in the range 300 pc to 2500 pc, so its shape over that range is the information. This happens to be sufficient to determine both the local ($z = 0$) volume mass density of the dark halo mass, which is about 0.01 M_o pc^{-3}, and the integral surface mass density of mass associated with the Galactic disk, which is 46 ± 9 M_o pc^{-2}. The K_z law is not determined reliably (here or elsewhere) at low z distances, so there is no determination yet possible of the local volume mass density of the disk mass, which is the slope of $K_z(z)$ near $z = 0$. At larger z the separability of the potential will not be valid, so the concept of K_z is not useful.

MARAN: What is the upper limit on dark matter in brown dwarfs in the disk, in solar masses per square parsec?

GILMORE: The upper limit on any form of dark mass is $\sim 2 \pm 12$ M_o pc^{-2}. For any old population of things with a velocity dispersion like that of old stars this corresponds to an undetectably small local mass density. Nevertheless, some brown dwarfs certainly exist. Just not enough of them exist to be important dynamically.

CHIU: Would you please comment on the large discrepancy between the observed luminosity to mass ratio and the observed velocity curves of some nearby galaxies?

GILMORE: The M/L ratio of the Galactic disk near the Sun is about three, of which two comes from the stars and one from the gas.   The consequence of other galaxies' rotation curves is that other spiral galaxies also have dark halos, which are dynamically important even well inside the optically bright part of the galaxy.

GOULD: The center of your K - F contours has F about 0 which appears to be about two sigma off under the assumption that there are almost no distance errors, but I think would be well under one sigma if the distance error were set by the same number of stars as the velocity errors.   Doesn't this cause you to worry about the validity of your density law?

GILMORE: The distance error must be set by the number of stars used to derive the distances, not by some other number.   The density law is in excellent agreement with that derived in other absolute magnitude ranges towards the SGP, and in excellent agreement with totally independent determinations by other people towards the NGP.   It would be a remarkable coincidence indeed (even more than two sigma?) if every independent data set were randomly in error in just the same way.

SCHECHTER   The contours of equal likelihood in the preprint which I received from Kuijken and yourself are cut off in the diagram so that one cannot see the most likely value of your F parameter.   It looks to be very different from that given by the rotation constraint. 1) What were the maximum likelihood values of K and F?   2) Shouldn't we be concerned that the maximum likelihood values of K and F lie far from the rotation constraints?

GILMORE: The kinematic analysis has no physical meaning with a solution which is not required to be consistent with the large-scale potential.   That is the whole point of a joint solution of the Poisson and Boltzmann equations.   Hence the straight values of K and F derived without the rotation constraint have no physical meaning.   The fact that one does not see any evidence for the large scale mass distribution in the local kinematics simply means the scale height of the dark stuff is much larger than that of the disk.

LU: If the thin disk is all identified matter, then all the dark matter is in the halo.   Is this consistent with Vera Rubin's rotational curve which implies a lot of galactic mass is in the halo?   And what distance scale does this dark stuff have in the galaxy?

GILMORE: The analysis is required to fit both the local velocity data and the galactic rotation curve, and hence is consistent with the large scale mass distribution.   Roughly half the mass interior to the Sun is in a dark halo, which must be in a roughly round distribution.   There is no known scale length associated with the dark stuff, in spite of a lot of efforts to find one.

RATNATUNGA: 1) What is the velocity dispersion of the K dwarfs around 1.5 kpc? 2) Follow up: Where is the thick disk with a velocity dispersion of 40 km/s?

GILMORE: 1) The velocity dispersion of the data can be approximated crudely as the sum of 2 Gaussians, of dispersions 22 km/s and 45 km/s. 2) It seems to have a dispersion of 45 km/s.

RATNATUNGA: It is a non-trivial exercise to convert log likelihood into one sigma errors. It is best done by simulation rather than theoretical approximation.

GILMORE: It is an extremely non-trivial exercise to relate likelihoods to confidence limits, as it depends on the form of the distribution function. A standard deviation is not a very relevant concept in the real world. At most distances for our data the distribution function is roughly a sum of two Gaussians, so one can crudely relate likelihood to sigmas. There is very little point in doing that, however. It is better to use likelihoods and forget parametric numbers. We attempted to scale likelihoods into significance limits so as to be able to quote numbers which mean something obvious. However, due to the complex coupling between the various parts of the analysis, the error distribution on the answer is not a simple Gaussian.

STATLER: A couple of comments and a question. First, not even the statisticians are entirely clear on how to turn likelihood values into sigmas on model parameters. Second, Mike Shull and I are trying to do a "definitive" job on the gas distribution by combining 21 cm, IUE, Copernicus, and anything else we can get our hands on. The latest value for the surface density of the ISM I generated was 10 $\pm$3 or so M_o pc^{-2}, but that is likely to go up a little. Finally, can you explain just how random distance errors propagate through your analysis and are reflected in the final answer?

GILMORE: I agree entirely with the unsuitability of using a sigma. Likelihoods are better, but less familiar. Distance (and velocity) errors are convolved with the model distribution function before comparison with the data, and so are included in the likelihood analysis. The surface mass density in the ISM which you are working on seems to be in excellent agreement with that derived earlier by Kulkarni and Heiles, which is the value we adopted.

AN ALTERNATIVE TO MASSIVE BLACK HOLES IN GALACTIC CENTERS

Hong-Yee Chiu

Goddard Space Flight Center

ABSTRACT: It appears that there are two classes of 'black hole'
objects,   those with masses around several solar masses and those
greater than 10^6 $M_\odot$. The former objects are believed to be collapsed
stars and are related to various X-ray or γ-ray objects, while the
latter objects are associated with AGN's. The conspicuous absence of
objects which exhibit characters of black holes in the rather large
mass gap between a few $M_\odot$ and say, 10^6 $M_\odot$, strongly suggests that
perhaps the nature of massive 'black   holes' might be completely
different from stellar sized black holes.

Soliton stars are massive objects (from 10^6 to 10^{14} $M_\odot$) not
necessarily in black hole configurations. Particles accreted to a
soliton star may give up nearly their entire rest energy, producing
high energy radiation. In this sense the soliton star is a nearly 100%
efficient energy converter, converting the rest energy of accreted
matter into radiation, such as those observed in AGN's.

1. INTRODUCTION

Recently Lee (1987), Friedberg et al. (1987a,b) and Lee and Pang
(1987) proposed a new class of astronomical objects, called soliton
stars, whose masses are quite large and may range from 10^6 $M_\odot$ to 10^{14}
$M_\odot$. Soliton stars correspond to nontopological solutions of Einstein's
field equations.   Physically they are fermions bound inside a
degenerate vacuum caused by a coherent background of Higgs-type bosons.
Because the physical environment of soliton stars is quite different
from that of ordinary stars, a discussion will be made around the
concept of a degenerate vacuum and the underlying physics of soliton
stars.   Physical   properties of soliton stars, as well as their
observational aspects, will then be presented.

2. PHYSICAL ENVIRONMENT OF SOLITON STARS

An ordinary vacuum is defined to be a state of minimum energy. As
there is always some energy associated with the presence of particles,
an ordinary vacuum is usually understood to be a state completely
devoid of particles, including radiation. However, the properties of

A. G. Davis Philip and P. K. Lu (eds.)
The Gravitational Force Perpendicular to
the Galactic Plane   81 - 90
©   1989   L. Davis Press

even an ordinary vacuum can be quite complex. Since 1949 it has been known that vacuum can be polarized; this polarization has been observed as a shift of atomic energy levels (Lamb shift), which results from the polarization of the vacuum by the atomic electric field. In fact, if the electric field energy density $E^2/8\pi$ approaches $m_e c^2/(\hbar m_e c)^3$ [i.e., one electron rest energy $m_e c^2$ per Compton wavelength volume $\hbar/m_e c)^3$], the polarization of the vacuum can become so strong so as to create electron-positron pairs. The notion that the dielectric constant of a vacuum is unity is valid only in the extremely weak field limit.

According to particle physics theories, other types of vacua may also exist. One such vacuum, which hereafter will be referred to as the 'degenerate vacuum', is made of a coherent background of Higgs type bosons (see references in Lee 1987) . Despite many compelling theoretical reasons that Higgs type bosons should exist, so far it has not been demonstrated and presumably the failure for the discovery is due to their rather large mass (> 30 m_p, m_p - the proton mass). Higgs type bosons have the property that they can modify the observed masses of other strongly interacting particles. The ability of Higgs type bosons to modify masses of particles can be likened to the modification of the dielectric properties of the ordinary vacuum by an electric field.

The degenerate vacuum has been discussed in conjunction with the inflationary phase during the early evolution of our Universe. According to this theory, during the early epoch of the universe $\leq 10^{-36}$ sec right after creation, the Universe was dominated by the presence of a degenerate vacuum and it was during this stage the Universe acquired its properties of isotropy and homogeneity. After this epoch the degenerate vacuum decayed and all observed matter is the decay product of this degenerate vacuum.

3. DEGENERATE VACUUM AND SOLITON STARS

So far many properties of Higgs type bosons remain unknown, although as we mentioned earlier, there are compelling reasons to believe that they not only exist but also play fundamental roles in the theory of elementary particles. Despite the uncertainties surrounding the Higgs type bosons, it is still possible to incorporate the Higgs type boson field in the theory of relativity from which the most general form of structure equations of self gravitating objects are derived. To achieve this one needs an expression of the self interaction energy of a Hermitian scalar field σ (Higgs type boson field) in addition to a fermion field φ. Let the mass associated with the Higgs type boson be μ. The simplest example of the self-interaction $U(\sigma)$ is of the following form (for convenience, the system of units used here is one in which $\hbar = c = 1$):

$$U(\sigma) = \tfrac{1}{2} \mu^2 \sigma^2 \left(1 - \frac{\sigma}{\sigma_0}\right)^2 \tag{1}$$

where σ_0 is a constant, such that for the ordinary vacuum $\sigma = 0$ while for the degenerate vacuum $\sigma = \sigma_0$.

Consider a system of fermions embedded in a sphere of radius R whose interior is characterized by the state $\sigma = \sigma_0$ (outside the sphere $\sigma = 0$). The mass of the fermions will be taken to be zero (as modified by the Higgs type boson field). The energy of the fermions, E_k, is given by:

$$E_k = \tfrac{1}{2} \left(\tfrac{3}{2}\right)^{5/3} \pi^{1/3} N^{4/3} /R \qquad (2)$$

Eq. (1) shows that the self-interaction energy vanishes both inside and outside the sphere. The thickness of the shell in which $U(\sigma)$ is not zero is of the order of the Compton wavelength of the Higgs type boson, i.e. μ^{-1} (in the system of units used). The associated interaction energy is concentrated on the surface and is:

$$E_s = 4 \pi \left(\tfrac{1}{6} \mu \sigma_0^2\right) R^2 \qquad (3)$$

The 'surface tension' coefficients due to the self-interaction is:

$$s \approx \tfrac{1}{6} \mu \sigma_0^2 \left[= \tfrac{1}{6} \mu c^2 (\sigma_0/\hbar c)^2 \approx 10^{24} (\mu\sigma_0^2/m_p^3) \; erg \; cm^{-3}\right] \qquad (4)$$

which is more than 10^{24} greater than that for water. Neglecting the self gravitational energy, an equilibrium requires a minimization of the sum $E_k + E_s$ (with respect to R), resulting in the condition

$$E_k = 2 E_s \qquad (5)$$

The total energy of the system E, the binding energy E_b, and the mass of the system M (the gravitational mass observed at large distances) are:

$$E = E_k + E_s = 3 E_s = 12 \pi s R^2 \approx s^{1/3} N^{8/9} \qquad (6)$$
$$M = E/c^2 \qquad (7)$$
$$E_b = k N^{8/9} m - Nm \qquad (8)$$

where m is the mass of the fermion before soliton binding and k is a constant. For any reasonably large value of N, E_b is always close to $-Nm$. A soliton star is thus a tightly bound object. This large binding energy also confirms our earlier assertion that inside soliton stars the rest mass of particles are modified to become nearly zero. Essentially the soliton binding is achieved through the surface tension generated at the boundary of the coherent background Higgs type boson field.

The inclusion of the gravitational energy will add a negative contribution E_g:

$$E_g \approx - G M^2 / R \tag{9}$$

For large enough mass M, E_g eventually will dominate over the surface energy (3) which is proportional to M, resulting in the formation of a black hole. However, a numerical evaluation shows that the mass M at which a black hole configuration results is around 10^{14} $M_\odot$, with a radius of the order of a light year. Thus even soliton stars which are not black holes, have very large masses compared to other gravitationally bound objects.

4. BASIC FORMULATIONS

The equations of structure of soliton stars may be obtained from the Einstein field equations, when the Higgs type boson field is properly included. Conservation laws may be invoked to obtain the macroscopic equations for the coherent Higgs type boson field.

The Einstein field equations are:

$$\mathcal{G}_{\mu\nu} \equiv \mathfrak{R}_{\mu\nu} - \tfrac{1}{2} g_{\mu\nu} \, \mathfrak{R} = 8\pi \, G \, \mathfrak{T}_{\mu\nu} \tag{10}$$

where $\mathfrak{R}_{\mu\nu}$ is the Ricci curvature tensor and $\mathfrak{R} = g^{\mu\nu} \mathfrak{R}_{\mu\nu}$ is the scalar tensor, $\mathfrak{T}_{\mu\nu}$ is the stress energy tensor, which includes the effects of the Higgs field. In a spherical coordinate system with the line element ds given by:

$$ds^2 = -e^{2u} dt^2 + e^{2\bar{v}} d\rho^2 + \rho^2 (d\alpha^2 + \sin^2\alpha \; d\beta^2) \tag{11}$$

where u, v are time and spatial metric, and ρ, α and β are standard polar coordinate variables. The stress energy tensors are:

$$\mathfrak{T}_t{}^t \; = W + V - U \tag{12}$$
$$\mathfrak{T}_\rho{}^\rho \; = T + V - U \tag{13}$$
$$\mathfrak{T}_\alpha{}^\alpha = \mathfrak{T}_\beta{}^\beta = T - V + U \tag{14}$$

where W is the energy density of particles, T is the pressure, V and U are field quantities associated with the Higgs bosons (U, V are analogues of the potential and kinetic energies). The expressions for W and T are:

$$W = \frac{2}{8\pi^2} \int d^3p \; n_p \; \epsilon_p \tag{15}$$

$$T = \frac{2}{8\pi^2} \int d^3p \; n_p \; \epsilon_p \; \frac{p^2}{3 \, \epsilon_p} \tag{16}$$

The expression for U is given in Eq. (1) and that for V is:

$$V = \tfrac{1}{2} \, e^{-2\bar{v}} \left(\frac{d\sigma}{d\rho}\right)^2 \tag{17}$$

The tensor $\mathcal{G}_{\mu\nu}$ satisfies the Bianchi identity:

$$\mathcal{G}^{\mu}{}_{\nu;\mu} = 0 \tag{18}$$

and the subscript ($_{\nu;\mu}$) denotes covariant differentiation with respect to the coordinate μ. Applying covariant differentiation to the field equation (10), we obtain the conservation law which must be satisfied by the stress energy tensor $\mathcal{I}_{\mu\nu}$:

$$\mathcal{I}^{\mu}{}_{\nu;\mu} = 0 \tag{19}$$

W and T already satisfy conservation laws as in Eq. (19). The conservation law (19), when applied to U and V yields the field equations of σ:

$$e^{-2\bar{v}}\Big[\frac{d^2\sigma}{d\rho^2} + \Big(\frac{2}{\rho} + \frac{du}{d\rho} - \frac{d\bar{v}}{d\rho}\Big)\frac{d\sigma}{d\rho}\Big] + fS - \frac{dU}{d\sigma} = 0 \tag{20}$$

where f is the coupling constant between the fermion and the Higgs boson, so that the interaction between and the fermion (whose wave function is φ) is $f\bar{\varphi}\varphi$, and S is a quantity related to T, W:

$$S = \frac{2}{8\pi^2}\int d^3p\, n_p\, \epsilon_p^{-1}\,(m - f\sigma) \tag{21}$$

Because of the form of U, a solution of Eq. (20) is

$$\sigma = \sigma_0. \tag{22}$$

Indeed, it has been shown by Lee and Pang (1987) (hereafter referred to as LP) that the deviation of σ from σ_o is extremely small (around one part in 10^{17}). As a result, the effective mass $m^* = m - f\sigma$ is nearly zero:

$$m^* = m - f\sigma_0 \approx 0 \tag{23}$$

In the following m^* will be taken to be strictly zero.

Using Eq. (23), the structural equations of the soliton stars become:

$$\rho^2\mathcal{G}^t{}_t = e^{-2\bar{v}} - 1 - 2e^{-2\bar{v}}\rho\frac{d\bar{v}}{d\rho} = -8\pi G\rho^2(W + V + U) \tag{24}$$

$$\rho^2\mathcal{G}^{\rho}{}_{\rho} = e^{-2\bar{v}} - 1 + 2e^{-2\bar{v}}\rho\frac{du}{d\rho} = 8\pi G\rho^2(T + V - U) \tag{25}$$

which can be solved to yield the structure of soliton stars. Details of solutions for completely degenerate fermi gas configurations at zero temperature have been described by LP. Note that these equations are perfectly general. When specific assumptions on W and T are imposed, one obtains different types of soliton stars.

Suffice it to say that the solutions obtained by LP show that soliton stars have very large mass, radius, and very large number of particles. Order of magnitude wise, the mass M, the radius R, the particle number N are:

$$M \approx \hbar^2 c^2 / G^2 m \sigma_0^2 \approx 10^{45} \; (30 \, m_P/m)(30 \, m_P/\sigma_0)^2 \; g \tag{26}$$

$$R \approx \hbar^2 / Gm\sigma_0^2 \approx 10^{18} \; (30 \, m_P/m)(30 \, m_P/\sigma_0)^2 \; cm \tag{27}$$

$$N \approx (\hbar c)^{9/4} / G^{9/4} m^{3/2} \sigma_0^3 \approx 10^{76} (30 \, m_P/n)^{3/2} (30 \, m_P/\sigma_0)^3 \tag{28}$$

Note that M is of the order of 10^{11} solar masses, R is of the order of a light year, and N is thirty orders of magnitude greater than the baryon numbers of most galaxies.

5. ORIGIN AND EARLY EVOLUTION OF SOLITON STARS

To summarize, soliton binding is very different from that in ordinary stars, where gravitational force is the dominant source of binding. Soliton stars are bound by the surface tension found at the interface between the degenerate vacuum and the ordinary vacuum, and the importance of gravitational force is usually secondary (unless the mass of the soliton star is very large). It appears unlikely that soliton stars can be formed in gravitational collapse processes. Most likely, soliton stars are created with the Universe after the inflationary epoch, as a result of incomplete decay of the prevailing degenerate vacuum. A theory accounting for the genesis of soliton stars in a second-order transition in the early Universe has been proposed (Chiu 1989), and it has been found that for a large range of parameters, non-topological solitons can be cosmologically significant, contributing a significant fraction of the present mass density of the Universe. This conclusion is supported by our study which shows that a large amount of matter may have been accreted onto soliton stars during early evolution of the Universe.

Independent of the origin of soliton stars, the main purpose of this work is to explore the observational consequences of soliton stars, if they exist. These observational consequences can be compared with current or future observations. Assuming the existence of soliton stars, we now explore their composition and early evolution properties. As shown by Lee and Pang (1987), the mass of hadrons inside a soliton star would be modified by the Higgs field to be nearly zero. In a bound system the binding energy is proportional to the mass of the constituent particles; if the mass of the constituent particles vanishes, the binding energy will vanish and the bound state will disappear. An analogy is found in the binding of an electron in an atom: the binding energy of the K-shell electrons is roughly $\frac{1}{2}\alpha^2 m_e c^2$ where α is the fine structure constant and m_e is the electron mass. When m_e vanishes the binding energy also vanishes. In this paper we assume that protons are composed of quarks, and that quarks are the most fundamental particle species in nature.

Under this assumption, a proton crossing from an ordinary vacuum into an abnormal vacuum will disintegrate into quarks of nearly zero effective mass, and likewise, a suitable combination of quarks (with suitable energies) can recombine into a proton and migrate from the degenerate vacuum into the ordinary vacuum. The equilibrium between protons and quarks across the interface between an ordinary vacuum and a degenerate vacuum may be considered as a reversible reaction:

$$p(outside) \leftrightarrow p'(inside) + \Delta E_b, \quad \Delta E_b \approx m_p c^2 \tag{29}$$

$$p' \leftrightarrow 2u + d - \Delta E_q', \quad \Delta E_q' \approx 0, \tag{30}$$

where $\Delta E_q'$ is the binding energy of the proton with respect to its quark constituents inside a soliton star, and is very close to zero.

The equations that govern the equilibrium between a proton outside a soliton star and quarks inside are:

$$\mu_p + m_p^2 c^2 = 2\mu_u + \mu_d \tag{31}$$

where μ is the chemical potential for the particle specified. Defining $U_0 \equiv m_p c^2/kT$, the expression for μ for a nondegenerate gas of the particle species a is given by:

$$N_a = N_0 \, exp(\mu_a/kT) \, U_0^{-3} \, H(U_0) \, exp \, -U_0, \quad N_0 = (m_p c)^3/\pi^2 \hbar^3 \tag{32}$$

and $H(s)$ is a function given by:

$$H(s) = \int_0^\infty t^2 \, exp \, -\left[\sqrt{t^2 + s^2} - s\right] \cdot dt \tag{33}$$

and we have:

$$\begin{array}{llll} H(s) \to 2, & s \to 0, \, T \to \infty & \text{(relativistic)} \\ H(s) \to \sqrt{\frac{\pi}{4}} \, s^{3/2}, & s \to \infty, \, T \to 0 & \text{(nonrelativistic)} \end{array} \tag{34}$$

Eqs. (31) and (32) then yield a relation between the equilibrium compositions of p, u and d:

$$N_p/N_d^3 \approx N_0^{-2} \, exp \, (-U_0) \, U_0^{7.5} \tag{35}$$

Since the effective mass of the quarks is nearly zero, their number densities are roughly given by the black body radiation law, i.e.,

$$N_u \approx 1.8 \, N_0 \, U_0^{-3} \approx 0.3 \, aT^4/kT \tag{36}$$

Then we have (the superscript (e) is added here to denote equilibrium composition):

$$N_p^{(e)} \approx N_0 \, U_0^{-1.5} \, exp \, -U_0 \tag{37}$$

If the actual proton number density N_p is less than the equilibrium density, $N_p^{(e)}$, then an equilibrium configuration will require all available quarks u and d to revert to protons, and vice versa.

The proton number density in our universe can be obtained from cosmological models. Despite uncertainties in cosmological models, the proton number densities, N_p, in our universe in relation to the temperature can be estimated to within a factor of five from the following equation (which is obtained from a cosmological model that yields the present densities of particles and temperature):

$$N_P \approx 10^{-7} \ T^3, \qquad T \approx 10^{10} \ t^{-1/2} \tag{38}$$

Numerically $N_p^{(e)} > N_p$ at temperatures $T_r = 4 \times 10^{11}$ K. On account of the strong exponential factor of Eq. (37), this crossing temperature is rather insensitive to details of cosmological models. Note that the above treatment applies to all particle species which are composed of quarks.

We thus conclude that the equilibrium configuration prohibits the entry of protons (and in fact, other hadrons as well) into soliton stars during early epochs when the temperature of the universe is $\geq T_c$. During this stage soliton stars expand with the universe and their interior is composed of quark pairs in equilibrium with radiation. The temperature of soliton stars is essentially the same as the universe. However, at temperatures below T_c, the equilibrium configuration is reversed and protons (and other hadrons) can enter a soliton star. The rest energy of hadrons is converted into thermal energy and the temperature of the soliton star can be different from the universe. In fact, this heating can drastically increase the temperature of soliton stars. It is also conceivable that a large number of baryons in our universe could have been irreversibly locked inside soliton stars and their rest energy is almost totally converted into radiation.

6. ENERGETICS AND RADIATIONS

Due to the smallness of the effective mass of the quarks, their 'classical electron radius' (inside soliton stars) r_q is very large. Indeed, from the mass to particle number ratio of soliton star models it may be concluded that the effective mass m^* is at least 100 times smaller than the electron mass, m_e. This causes the opacity of quark matter inside soliton stars to be very large. In normal stars a very large opacity will reduce luminosity. However, since the effective mass m^* is small, the pair creation temperature $T_p = m^* c^2/\kappa$ is also small and at temperatures $T \gg T_p$ the internal energy is chiefly in the form of quark pairs in equilibrium with radiation. The annihilation radiation from the surface within one optical depth may escape and be radiated away, independent of opacity. The energy source of the emitted radiation is contained within the photosphere. This radiation mechanism is very different from that of ordinary stars, where the energy source is in the deep interior and through radiation transfer or convection the thermal energy is brought to the surface and radiated away. This radiation mechanism makes the radiation rate of soliton stars independent of opacity, only dependent on the available source at the surface. Indeed, the photosphere literally radiates itself away. As a good approximation, the surface temperature of a soliton star may be assumed to be the same as its interior.

Let T be the temperature of the soliton star, R be its radius, and M be its mass, then the luminosity L (up to Eddington's limit) is given by:

$$L = 4\pi \, \sigma \, T^4 \, R^2 \tag{39}$$

Since the radiated energy comes from the mass energy of the star, the mass M decreases with time. As mass decreases, the characteristics (such as R and T) also change with time. An evolutionary sequence can be constructed once the initial conditions are given.

7. MODEL CALCULATIONS

Although the work of LP deals with zero temperature fermion soliton stars, it can be easily extended to apply to nonzero temperature configurations. However, irrealistic model protons entering a soliton star must be accompanied by an equal number of electrons. While the mass of the proton is modified by the Higgs type boson field, the mass of the electron is not. Indeed, electrons may be the most massive particle inside such soliton stars. Solutions of somewhat different character from that of LP have been obtained (Chiu 1989).

However, at sufficiently high temperature when quark pairs dominate, the work of LP is still applicable. This condition is fulfilled for soliton stars whose temperatures are much less than the electron pair creation temperature ($\sim 7 \times 10^9$ K), so that electron pairs will not contribute much to the mass, while the temperature is still high enough so that the mass energy of the electrons are still small compared with those of quark pairs.

Using the model of LP, we have computed the evolutionary behavior of soliton stars. The lifetime for radiation appears to be rather short. At the end of evolution, electrons inside soliton stars should assume importance. Since the effective rest mass of the electron inside a soliton star is the same as its mass in ordinary vacuum, black hole configuration is possible for much smaller soliton star masses.

8. DISCUSSION

In this paper we discussed the possible evolution of soliton

stars.   Although many of their properties remain unknown, it is
possible that they may fulfill the role of dense massive nuclei of
AGN's.   Further work on other soliton star configuration based on
different assumptions about the self interaction energy $U(\sigma)$ should
reveal many more interesting properties.   In addition, the role played
by electrons should be explored within the framework of the theory of
LP.

ACKNOWLEDGEMENTS.

    I would like to thank Dr. Phillip Lu for his excellent
organizational work on the K_z meeting in which this work is reported.
The very generous support of the Laboratory for Atmospheres for this
work is greatly acknowledged, especially that of Drs. E. Maier and M.
Geller.

REFERENCES

Chiu, H. Y.   1989 Astrophys. J., submitted.
Chiu, H. Y.   1989, in preparation.
Friedberg, R., Lee, T. D. and Pang, Y. 1987   Phys. Rev. D 35,
    3640.
Friedberg, R, Lee, T. D. and Pang, Y. 1987 Phys. Rev. D 35, 3658.
Frieman, J. A., Gelmini, J. B. Gleiser, M. and Kolb, E. W. 1988
    Phys. Rev. Ltrs. 60, 2101.
Lee, T. D. 1987 Phys. Rev. D 35, 3637.
Lee, T. D. and Pang, Y. 1987 Phys. Rev. D 35, 3678.

DISCUSSION

GOULD: How many particles (protons) are there in an average soliton
star of a mass, say 10^{14} solar masses?

CHIU: The number of particles (protons) is around 10^{76}.   Before
soliton binding this would amount to 10^{19} solar masses of protons.
However, the coherent Higgs field modifies the proton mass.   The proton
mass inside a soliton star is much smaller, being in the neighborhood
of 10^{-6} m_p.   This means that during formation virtually all the rest
mass of the 10^{19} solar masses has been converted to energy, most likely
radiation.

MARAN: Do these still exist in the current epoch and if so, if one
were present in a cluster of galaxies would it interact
gravitationally?

CHIU: Yes, their gravitational behavior is not modified.

A SEARCH FOR DISTANT STARS IN THE GALACTIC HALO AND THICK DISK

Ken Croswell

Harvard-Smithsonian Center for Astrophysics

Stars in the Galactic halo and thick disk are the oldest stars in the Galaxy. They, therefore, give us great insight into the origin and early evolution of the Milky Way Galaxy. This project is an attempt to discover distant stars embedded in the halo and thick disk, stars that lie many kiloparsecs above the galactic plane. Goals include the measurement of the halo's line-of-sight velocity dispersion as a function of distance from the galactic plane, an examination of whether or not the halo has an abundance gradient, and a determination of the parameters of the thick disk. We can also use high-velocity stars at large distances from the galactic center to put a lower limit on the Galaxy's mass, provided we assume the stars are bound to the Galaxy.

In addition, a search for distant halo and thick disk stars will be much less biased than solar neighborhood studies of the halo and thick disk. Since the Sun lies in the middle of the thin disk, the vast majority of nearby stars are members of the thin disk population. To pick out the few halo and thick disk stars from the nearby star population, one must select stars either on the basis of high velocity relative to the Sun (e. g., Eggen, Lynden-Bell, and Sandage 1962; Sandage and Kowal 1986; Fouts and Sandage 1986; Carney and Latham 1987) or on the basis of low metallicity (e. g., Bond 1980; Norris, Bessell, and Pickles 1985; Beers, Preston, and Shectman 1985). If instead we look at faint, distant stars far above the galactic plane, we can obtain a sample of halo and thick disk stars that is kinematically unbiased and only slightly metallicity biased. (A slight metallicity bias arises because a star's luminosity depends on its metallicity: metal-poor dwarfs are less luminous than metal-rich dwarfs, and metal-poor giants are more luminous than metal-rich giants.)

We have surveyed one square degree of Selected Area 57, near the North Galactic Pole. Our survey includes all stars in this square degree that have V magnitudes between 13.5 and 17.8 mag. and whose colors are bluer than (B - V) = 1.0. There are 247 stars in all. We obtained spectra of these stars at the Multiple Mirror Telescope and Strömgren photometry of about 150 of them at Kitt Peak and San Pedro Martir. Our goal is to determine velocities, metallicities, luminosities, and distances for the stars in the survey.

A. G. Davis Philip and P. K. Lu (eds.)
The Gravitational Force Perpendicular to
the Galactic Plane   91 - 94
©   1989   L. Davis Press

As we rise farther and farther above the galactic plane, we expect that we will sample different populations: first the thin disk (σ_w = 20 km/s), then the thick disk (σ_w = 40 km/s) and finally the halo (σ_w = 80 km/s). The velocities we have measured for our 247 stars confirm this expectation.

TABLE I

Velocity Dispersion As A Function Of Apparent Magnitude

V Magnitude	Number of Stars	σ_w (km/s)
13 to 14	20	25 $\pm$6
14 to 16	101	47 $\pm$5
16 to 18	126	76 $\pm$7
TOTAL	247	62 $\pm$4

The brightest (13 mag $\leq$ V $\leq$ 14 mag) and presumably nearest stars have a velocity dispersion similar to that of the thin disk. Stars of intermediate brightness (14 mag $\leq$ V $\leq$ 16 mag) have a velocity dispersion comparable to that of the thick disk. And the faintest (16 mag $\leq$ V $\leq$ 18 mag) and presumably farthest stars have a velocity dispersion like that of the halo.

The total velocity dispersion of the sample is quite high (σ = 62 $\pm$4 km/s), about halfway between that of the thick disk and that of the halo. This high velocity dispersion suggests that most of our stars are indeed members of the thick disk and halo, even though we did not pick out stars on the basis of either kinematics or metallicity.

ACKNOWLEDGEMENTS

This project would have been impossible without the great help of Dave Latham, Bruce Carney, Bill Schuster and Luis Aguilar.

REFERENCES

Beers, T. C., Preston, G. W. and Shectman, S. A. 1985 Astron. J. 90, 2089.
Bond, H. E. 1980 Astrophys. J. Suppl. 44, 517.
Carney, B. W. and Latham, D. W. 1987 Astron. J. 93, 116.
Eggen, O. J., Lynden-Bell, D. and Sandage, A. R. 1962 Astrophys.

J. 136, 748.
Fouts, G. and Sandage, A. R. 1986 Astron. J. 91, 1189.
Norris, J., Bessell, M. S. and Pickles, A. J. 1985 Astrophys. J.
    Suppl. 58, 463.
Sandage, A. R. and Kowal, C. 1986 Astron. J. 91 , 1140.

DISCUSSION

RATNATUNGA: Was the sample of 247 stars selected on the basis of photographic photometry, and what is the error in that photometry? How many of those 247 stars with (B-V) < 1 do you expect to be more than 10 kpc from the Galactic plane?

CROSWELL: The sample was selected on the basis of photographic RGU photometry from the Basel group. The error in that photometry is probably 5 to 10%. I do not yet know how many stars we have with distances greater than 10 kpc, maybe a couple dozen or so.

LEE: From observation the dispersions in radial velocities are, and correspond to:

V 13 - 14	25 ±6	===>	Old Disk Stars
14 - 16	47 ±5	===>	Thick Disk Stars
16 - 18	76 ±7	===>	Halo Stars

Is there any exception? Is there any star that has a bright magnitude but high sigma?

CROSWELL: Yes, there are exceptions. Surely there are some halo stars with 14 < V < 16 and some thick disk stars with 16 < V < 18. But the general trend is that the brightest stars are thin disk stars, the intermediate stars are thick disk stars, and the faintest stars are halo stars.

WEIS: May I assume that you discovered no yellow degenerate stars?

CROSWELL: Correct.

DA COSTA: Selecting only stars with (B-V) < 1.0 will not only remove faint K dwarfs but also highly luminous K giants that will be far into the halo. These would be the most interesting stars for constraining the galaxy mass.

CROSWELL: K giants would indeed be very interesting, but the vast majority of the K stars would prove to be dwarfs.

KING: Besides merely using the escape velocity, one can use the velocity dispersion in a virial relationship, to put restrictions on the potential. After all, the Little and Tremaine study of globular clusters and dwarf spheroidal galaxies has only about a dozen objects. I hope that you will also have a chance to check for duplicity.

CROSWELL: Unfortunately we have only one spectrum of each star.

A PROGRAM TO DETERMINE K_z USING FAINT DWARF F-STARS AT THE
SOUTH GALACTIC POLE I. PRELIMINARY RESULTS

Phillip K. Lu[*]

Center for Galactic Astronomy
Western Connecticut State University and
Van Vleck Observatory, Wesleyan University

ABSTRACT: A study is in progress to determine the galactic
gravitational force law K_z, and the local mass density using faint
dwarf F-stars as tracer objects. The primary objective of this
investigation is to search for evidence of the thick disk and to make a
dynamical study of the missing mass and dark matter in the disk of the
Milky Way Galaxy. Objective-prism plates covering approximately 100
square degrees, centered on the South Galactic Pole, using UK Schmidt
films and Michigan Curtis Schmidt plates, have been examined.
Observations, using direct CCD photometry with uvby, $H\beta$ colors were
obtained for 109 stars. It is shown that about 6% of these F-stars are
subdwarfs. Of 257 spectra obtained, using a 2D-Frutti system at a
dispersion of 43 Å/mm, 95% are now identified as F-stars. Radial
velocities, obtained from these spectra using cross-correlation
techniques, yield an accuracy of about $\pm$ 10 km/s.

1. INTRODUCTION

A project to determine the galactic gravitational force law, K_z,
the local mass density and dark matter in the Milky Way Galaxy using
faint main-sequence F-stars as tracer objects to one kpc is now in
progress. The primary objectives are to investigate the velocity
distribution, to search for evidence of a thick disk and to make a
dynamical study of the missing mass and dark matter in the galactic
disk.

The population of faint dwarf F-stars at the South Galactic Pole

[*]Visiting Astronomer, Cerro Tololo Inter-American Observatory, National
Optical Astronomy Observatories, which is operated by the Association
of Universities for Research in Astronomy, Inc., under contract with
the National Science Foundation.

A. G. Davis Philip and P. K. Lu (eds.)
The Gravitational Force Perpendicular to
the Galactic Plane   95 - 104
○   1989   L. Davis Press

from the objective-prism spectral survey has been published (Lu 1988). Some results and the spectral classification using 2D-Frutti spectra obtained at the Cerro Tololo Inter-American Observatory (CTIO) were reported at the Third Conference in Astrophysics in Suchow, China (Lu 1989a,b). Preliminary results obtained with the direct CCD camera on uvby, Hβ photometry and other results are presented here. The determination of radial velocities, using the 2D-Frutti system is in progress and will be reported at a later date. These data are required for a dynamical investigation of the local mass density near the Sun.

The dynamical determination of the total mass density in the solar neighborhood by means of the stellar velocity distribution has been recognized as a very difficult task. Some of the early investigations of this problem were made by Oort (1932, 1965) and led to the first suggestion of missing mass in the Galaxy. The nature of the missing mass has been attributed to dark matter which includes low luminosity and low mass stars, brown and white dwarfs, black holes and possibly quarks.

Dynamical studies, using giants and nearby A- and F-stars, have suggested that about half the total matter density near the Sun is unaccounted for (Bahcall 1984a,b,c, Bahcall and Casertano 1985, Bahcall, Hut and Tremaine 1985). It is not clear, however, whether this discrepancy is a localized phenomenon confined to the solar neighborhood or is a general property of the Milky Way Galaxy, although this invisible mass must be in a disk with a scale height not exceeding 700 pc (Bahcall 1984b). Gilmore and Kuijken (1989) have shown there is little evidence of a thick disk.

The distribution of mass density and stars perpendicular to the galactic disk is generally determined by comparing observational data with the results from a theoretical model assuming an ellipsoidal mass distribution. In recent years, two new methods of analyzing tracer data have been proposed (Bahcall 1984a,b,c; Kuijken and Gilmore 1989). These two proposed methods of tracer analysis are different in approach. On the one hand, Bahcall's method compares the observed tracer density profile to various trial density profiles, each corresponding to a particular trial potential, and solves both the Poisson equation, $\nabla^2 \Phi = 4\pi G[\rho_{disk} + \rho_{halo}]$ and the first moment of the Boltzmann equation, that describe the motion of an isothermal population of stars perpendicular to an axisymmetric galactic disk simultaneously. In Bahcall's method, the solution depends upon the ratio of the effective halo mass density to the mass density in the plane of the disk.

Kuijken and Gilmore (1989), on the other hand, have proposed a much simpler technique. They concentrate on determining the linear term in the high altitude potential, using high-latitude tracers, thus, a specific population of stars at the galactic poles. This term is proportional to the column density of the disk. This measured column density may then be related to what the column density would be if there were no missing mass.

In Bahcall's case, the determination of a distribution function and a gravitational potential is based on the joint solution of the Boltzmann and Poisson equations (Bahcall 1984a,b,c). The total mass density in the solar neighborhood is found to equal 0.185 $\pm$0.02 $M_\odot/pc^3$ according to Bahcall. A dynamical determination of the local mass density using A- and F-stars in the region of the North Galactic Pole, based on radial velocities and photometric data (Hill, Hilditch and Barnes 1979; hereafter referred to as HHB) and a theoretical model due to Camm (1950, 1952) yield a value of 0.14 $M_\odot/pc^3$. The total mass density of known matter in the solar neighborhood is found to be 0.108 $M_\odot/pc^3$ by HHB. However, Bahcall (1984b) has pointed out that HHB's mass density requires a small but uncertain correction due to the existence of binaries (Bahcall et al. 1985). Therefore, there is no large difference between HHB and Bahcall's results.

In another aspect, Gilmore and Reid (1983) argued that faint star counts could not be modeled by the halo and the disk two-component system, which was disputed by Bahcall and Soneira (1984). Using magnitude-limited velocity and composition data in the direction of the polar caps (Gilmore and Wyse 1985) and magnitude-limited proper motion data (Wyse and Gilmore 1986) Gilmore and Wyse reported additional evidence of a third population in the Galaxy. In all cases, this third population has been referred as the thick disk. Carney et al. (1989) have further suggested that this so-called third population has an age like that of the disk globular clusters, such as 47 Tuc.

Regardless of what methods of stellar dynamics are used, a specific type or population of stars to a z-distance of 1.5 kpc is necessary for the investigation of the velocity dispersion. In the presently available data, little information is known to a z-distance beyond one kpc for the unevolved main-sequence F-stars. In most cases, large numbers of these stars are needed for the theoretical models. However, in a recent study, Gould (1989) showed that a large sample of tracer objects is important for the statistical analysis. The effectiveness of a survey deteriorates greatly the more the tracers deviate from being isothermal. In another word, a few hundred tracers are sufficient to determine whether the disk has a substantial quantity of missing matter.

The first large-scale attempt by Oort (1932), about 50 years ago, was partially successful in defining the galactic force law analytically. More recent work has been carried out by Hill (1960) and by Oort (1960). The sample of stars available for Oort's work was small and consisted of inhomogeneous groups of old disk stars. It is remarkable that Oort was able to draw such fundamentally sound conclusions from such fragmentary data. Since the time of Oort's study, a great deal of observational material has been gathered concerning the kinematics and physical properties of the high-velocity stars by Fricke (1949), Eggen (1962, 1964) and Sandage (1969).

The advantages of using F-stars at the galactic poles as compared with any other stars have been pointed out by Upgren (1977), King (1983) and Freeman (1985). First, at high galactic latitudes, the density distribution of a group of stars in the z-direction related to its velocity distribution (defined as W velocity) has less effect due to the rotation of the galaxy. Second, F-stars are reasonably abundant (about 64 stars per square degree between 10 to 16.5 magnitude, (Lu 1988), which is comparable to Bok and Basinski (1964)). F-stars would also be identifiable at great distance from the disk. However, King (1989) has recently pointed out that early F-stars may not be in dynamical equilibrium.

Measuring K_z directly has never been very satisfactory. The primary reason for this lies in the shortage of radial velocity data for very faint and distant stars in the direction of the galactic poles. The importance of obtaining radial velocities of very faint and distant stars has been pointed out by many investigators (Schmidt 1965, Blaauw 1978, King 1983 and Freeman 1985). The radial velocities of very faint F-stars is vitally important for large scale galactic work and high precision is not necessary. Thus, an average velocity accuracy of ± 10 km/sec for magnitudes brighter than V = 16 is adequate for this purpose.

Projects to determine radial velocities of nearby dwarf A- and F-stars are many, notably those by Andersen and Nordström (1983a,b,c 1985), Hill et al. (1979) and McFadzean et al. (1982, 1987). Since these surveys have a limiting magnitude of about 9 or a distance of about 300 pc, they would not provide the necessary data to study the large distance and thick disk problem. All of Andersen and Nordström's stars were selected from the 4-color photometry of 4000 dwarf A and F stars brighter than 8.5 (Olsen 1983). Other investigators are studying very distant field halo stars in order to determine the kinematics and metal abundance distribution in the outer regions of the galactic halo based on late giants (Dessureau and Upgren 1975, Hartkopf and Yoss 1982, Ratnatunga and Freeman 1984, Andersen and Jensen 1985). Although both luminous and numerous at large distances, they are evolved objects; thus their absolute magnitudes are uncertain at large z-distances.

2. OBSERVATIONS AND RESULTS

Using a 1.5 degree prism with a dispersion of 1360 Å/mm, Blanco (1974) has found that the prism allows the recognition of A4 - G2 spectral types up to a photographic magnitude of 16.0. Although detailed spectral classifications are not normally possible at this dispersion, a broad range of spectral types are recognizable under good seeing conditions. A similar low-dispersion survey of the northern sky with a 1.8 degree prism, which provided a dispersion of approximately 1500 Å/mm centered at 4500 Å, was carried-out by Pesch and Sanduleak

(1983), and Sanduleak and Pesch (1984). With Eastman Kodak IIIa-J plates, baked in forming gas, the plates reach a limiting magnitude of B = 18.0 with a 75 minute exposure time.

Objective-prism plates centered at the South Galactic Pole have been obtained during December 14 - 22, 1987 and August 19 - 22, 1988 for the entire region of the 100 square degrees using the Michigan Curtis Schmidt telescope at CTIO. There are 9 objective-prism plates, and each has a 5 x 5 square-degree field, covering a total 100 square-degree area. Both 15 and 90 minute exposure times were used for each region to insure that all stars within the magnitude range of 8.5 to 16 were properly exposed. Using the thin prism and baked IIIa-J emulsion, the plates can readily reach B = 16.5 with a 90 minute exposure time and the spectral types in the broad range are easily identifiable (Lu 1988).

Objective-prism and direct film copies were also provided by the UK Schmidt Unit at the Royal Observatory, Edinburgh, Scotland. These films have superb quality, are of comparable dispersion, and each film covers a 6.3 by 6.3 square-degree field. Since the films were obtained in the search for extragalactic objects at the SGP and have a limiting magnitude of 21 with a 70 minutes exposure time, stars brighter than 15th magnitude would be generally overexposed.

In the case of the F-stars, the misidentification of a giant as a dwarf is not of great importance because there are very few F-type stars above the main sequence in this region of the HR diagram. However, a problem does exist with the F subdwarf stars of high ultraviolet excess. These old stars lie about a magnitude below the ZAMS, according to Sandage (1970), and can be expected to predominate in regions well away from the galactic plane. Distinguishing the subdwarfs from the normal dwarfs is nearly impossible using objective-prism spectra. However, photometrically, it is possible to identify them from their luminosity and metallicity using 4-color and H β photometry.

Direct CCD photometry was obtained during 7 nights at CTIO (December, 1987, August and November, 1988) for the newly identified F-stars using the 0.9 m telescope with a RCA 1 and a TI CCD chip. It normally requires 4 to 5 minutes integration time to reach a 16th magnitude star in the vby bandpasses. Since the RCA chips have very low quantum efficiency in the UV region 80% of the observing time would be used to measure the u-band. As for TI CCD chips, triple integrating times are needed in the u-band to achieve the same intensity.

Standard stars were selected from Crawford and Mander (1966), Crawford (1975), Crawford and Barnes (1970), Grönbeck and Olsen (1976), Grönbeck, Olsen and Strömgren (1976) and Twarog (1984). Since all stars in Crawford's lists are bright stars, a neutral density filter of 5 to 7.5 mag was used. The transmission curves of the neutral density filters vary, hence, corrections have to be applied when appropriate.

Twarog (1984) has provided a set of secondary uvby standard stars near
the South Galactic Poles. The reduction has shown that these fainter
standards appears to have a small systematic difference from Crawford
standards.

All instrumental magnitudes were obtained either using the SPHOT
package at CTIO or QPHOT within the APPHOT packages of KPNO.   The
DAOPHOT program developed by Stetson (1987) was used for a few crowded
fields near NGC 288. Air mass corrections were applied using the CTIO
standard extinction coefficients and the instrumental indices were
transferred to the standard system using the following relations:

$$V_{st} = A + B \ (b\text{-}y)_{st} + y_{ins}$$
$$H\beta_{st} = C + D \ H\beta_{ins}$$
$$(b\text{-}y)_{st} = E + F \ (b\text{-}y)_{ins}$$
$$m_{1st} = G + H \ m_{1ins} + I \ (b\text{-}y)_{st}$$
$$c_{1st} = J + K \ c_{1ins} + L \ (b\text{-}y)_{st}$$

where the subscripts (st) and (ins) refer to the specific index on the
standard system and the instrumental system, respectively.   The
distribution of magnitudes, colors and spectral types are shown in
histograms (Figs. 1a - 1f). Color-magnitude and color-color diagrams
are shown in Figs. 2a - 2c.

Spectral types of 257 F-star candidates have been obtained using
a 2D-Frutti system at a dispersion of 43 Å/mm at CTIO. The instrument
includes a 2-dimensional photon detector, a combination of image-tube
and CCD system.   It is possible to obtain spectroscopic and radial
velocity data using this instrument.   A Multiple Object Fiber Feed
Spectrograph, called ARGUS, may be also used.   Radial velocity
reductions using cross-correlation techniques within IRAF (Image
Reduction and Analysis Facility) is in progress. Of the 257 stars, 95%
are now identified as F-stars.

The 2DF spectral types relative to (b-y) determined in this study
are shown in Fig. 3.   The large scatter indicates that there may be
both abundance and luminosity effects.

We wish to thank Dr. Robert Williams and his staff at CTIO for
all their support, and to Dr. Sue Tritton at the Royal Observatory,
Edinburgh, Scotland, who has provided the needed film copies to conduct
our earlier search for F-stars.   Thanks are also due to Drs. John
Bahcall, Victor Blanco, William van Altena and Arthur Upgren, who have
served as project consultants.   This research has been supported in
part by grants from the Connecticut State University system and from
the National Science Foundation grant AST 8713183.

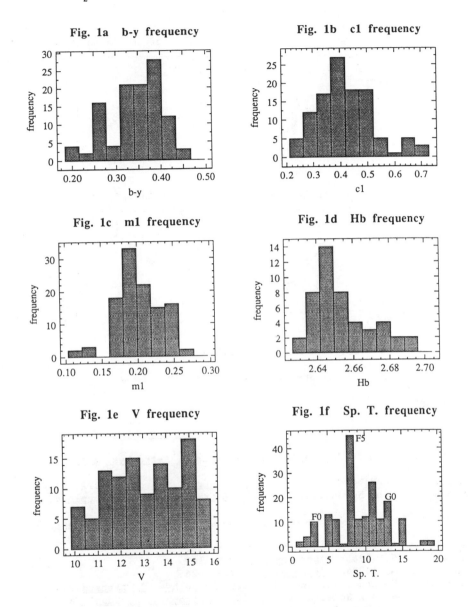

Fig. 1a   b-y frequency

Fig. 1b   c1 frequency

Fig. 1c   m1 frequency

Fig. 1d   Hb frequency

Fig. 1e   V frequency

Fig. 1f   Sp. T. frequency

Fig. 1.   Histograms of Colors, Magnitudes and Spectral Type.

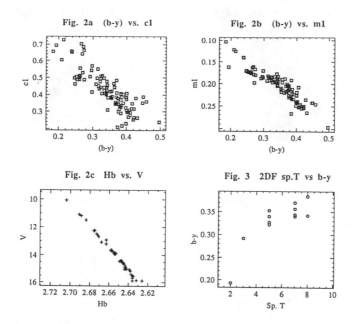

Fig. 2.   Color-Magnitude and Color-Color Diagrams.

Fig. 3.   2DF Spectral Type versus (b-y).

REFERENCES

Andersen, J. and Nordström, B.   1983a   <u>Astron. Astrophys. Suppl.</u>
    52, 471.

Andersen, J. and Nordström, B.   1983b   <u>Astron. Astrophys. Suppl.</u>
    52, 479.

Andersen, J. and Nordström, B.   1983c   <u>Astron. Astrophys.</u> **122**, 23.

Andersen, J. and Nordström, B.   1985   <u>Astron. Astrophys. Suppl.</u> **62**,
    355.

Andersen, T. B. and Jensen, K. S.   1985   <u>Astron. Astrophys. Suppl.</u> **59**,
    361.

Bahcall, J. N.   1984a   <u>Astrophys. J.</u> **276**, 156.

Bahcall, J. N.   1984b   <u>Astrophys. J.</u> **276**, 169.

Bahcall, J. N.   1984c   <u>Astrophys. J.</u> **287**, 926.

Bahcall, J. N. and Casertano, S.   1985   <u>Astrophys. J. Letters</u> **29S**, L 7.

Bahcall, J. N. and Soneira, R. M.   1984   <u>Astrophys. J. Suppl.</u> **55**, 67.

Bahcall, J. N., Hut, P. and Tremaine, S.   1985   <u>Astrophys. J.</u> **290**, 15.

Blaauw, A.   1978   in <u>Astronomical Papers Dedicated to Bengt Strömgren</u>,
    A. Reiz and T. Andersen eds., Copenhagen Univ. Obs. Press, p. 33.

Blanco, V.   1974   <u>Publ. Astron. Soc. Pacific</u> **86**, 841.

Bok, B. J. and Basinski, J.   1964   <u>Memoirs of Mt. Stromlo Obs.</u> **16**.

Camm, G. L.   1950   <u>Monthly Not. Roy. Astron. Soc.</u> **110**, 305.

Camm, G. L. 1952 Monthly Not. Roy. Astron. Soc. 112, 155.
Carney, B. W., Latham, D. W. and Laird, J. B. 1989 Astron. J. 97, 432.
Crawford, D. L. and Mander, J. 1966 Astron. J. 71, 114.
Crawford, D. L. 1975 Astron. J. 80, 955
Crawford, D. L. and Barnes, J. V. 1979 Astron. J. 75, 946.
Dessureau, R. L. and Upgren, A. R. 1975 Publ. Astron. Soc. Pacific 87, 737.
Eggen, O. J. 1962 Roy. Obs. Bull. 51, E 79.
Eggen, O. J. 1964 Roy. Obs. Bull. 84, E 111.
Freeman, K. C. 1985 in IAU Colloquium No. 88, Stellar Radial Velocities, A. G. D. Philip and D. W. Latham, eds., L. Davis Press,. p. 223.
Fricke, W. 1949 Astron. Nachr. 278, 49.
Gilmore, G. and Kuijken, K. 1989 in The Gravitational Force Perpendicular to the Galactic Plane, A. G. D. Philip and P. K. Lu, eds., L. Davis Press, Schenectady, p. 61.
Gilmore, G. and Reid, N. 1983 Monthly Not. Roy. Astron. Soc. 202, 1025.
Gilmore, G. and Wyse, R. F. G. 1985 Astron. J. 90, 2015.
Gould, A. 1989, preprint.
Grönbeck, B. and Olsen, E. H. 1976 Astron. Astrophys. Suppl. 25, 213.
Grönbeck, B., Olsen, E. H. and Strömgren, B. 1976 Astron. Astrophys. Suppl. 26, 155.
Hartkopf, W. I. and Yoss, K. M. 1982 Astron. J. 87, 1679.
Hill, E. R. 1960 Bull. Astron. Inst. Netherlands, 15, 1.
Hill, G., Hilditch, W. and Barnes, J. V. 1979 Monthly Not. Roy. Astron. Soc. 186, 813.
Jones, D. H. P. 1962 Roy. Obs. Bull. No. 52.
King, I. R. 1966 Astron. J. 71, 64.
King, I. R. 1983 in Kinematics, Dynamics, and Structure of the Milky Way, W. L. H. Shuter, ed., Astrophys. and Space Sci. Library, Vol. 100, Reidel, Dordrecht, p. 53.
King, I. R. 1989 in The Gravitational Force Perpendicular to the Galactic Plane, A. G. D. Philip and P. K. Lu, eds., L. Davis Press, Schenectady, p. 147.
Kuijken, K. and Gilmore, G. 1989 Monthly Not. Roy. Astron. Soc. 239, in press.
Lu, P. K. 1988 in Calibration of Stellar Ages, A. G. D. Philip, ed., L. Davis Press, Schenectady, p. 217.
Lu, P. K. 1989a in The 3rd. Conference on the Active Galactic Nuclei, Nov. 10-14, Suchow, China, in press.
Lu, P. K. 1989b Bull. Am. Astron. Soc. 21, 779.
McFadzean, A. D., Hilditch, R. W. and Hill, G. 1982 Monthly Not. Roy. Astron. Soc. 205, 525.
McFadzean, A. D., Hilditch, R. W., Hill, G., Aikman, G. C. L. and Fisher, W. A. 1987 Monthly Not. Roy. Astron. Soc. 224, 393.
Olsen, E. H. 1983 Astron. Astrophys. Suppl. 54, 55.
Oort, J. H. 1932 Bull. Astron. Insts. Netherlands 6, 249.

Oort, J. H.   1960   Bull. Astron. Insts. Netherlands 15, 45.
Oort, J. H.   1965   in Galactic Structure, A. Blaauw and M. Schmidt,
    eds., Univ. of Chicago Press, Chicago, p. 455.
Pesch, P. and Sanduleak, N.   1983   Astrophys. J. Suppl. 51, 171.
Ratnatunga, K. and Freeman, K. C.   1984   in Astronomy with Schmidt-Type
    Telescopes, M. Capaccioli, ed., Reidel, Dordrecht, p. 261.
Sandage, A. R.   1969   Astrophys. J. Suppl. 2, 195.
Sandage, A. R.   1970   Astrophys. J. 162, 841.
Sanduleak, N. and Pesch, P.   1984   Astrophys. J Suppl. 55, 517.
Schmidt, M.   1965   in Galactic Structure, A. Blaauw and M. Schmidt,
    eds., Univ. of Chicago Press, Chicago, p. 513.
Stetson, P. B.   1987   Publ. Astron. Soc. Pacific 99, 191.
Turon-Lacarrieu, C.   1971   Astron. Astrophys. 14, 95.
Twarog, B. A.   1984   Astron. J. 89, 523.
Upgren, A. R.   1977   in IAU Highlights of Astronomy, Vol 4. pt II,
    E. A. Müller, ed., Reidel, Dordrecht, p. 75.
Wyse, R. F. G. and Gilmore, G.   1986   Astron. J. 91, 855.

DISCUSSION

RATNATUNGA: The Bok photographic photometry is 0.02 mag rms or better.

LU: If that is the case, the scattering in the correlation may be due to the metal abundance in these stars.   Once we complete Strömgren 4-color observations, the abundance effect can be then eliminated.

YOSS: What is your 2D-frutti accuracy for spectra?   Are the spectral classifications based on visual inspection or from measurement of line ratios?

LU: All the spectral classifications are visually inspected at present.   Line ratios will be used in the final analyses.   The accuracy of the 2DF spectral types should be within 1-2 sub-class.   MK standards were obtained using the 2DF with the same dispersion and set-up.   This will minimize the classification error.

RATNATUNGA: By using digital spectra from PDS scans of the objective prism images one may be able to do better spectral classification.

LU: To PDS the objective-prism for the entire region is really not necessary since the 2DF has a higher dispersion and thus more accurate spectral classifications can be made.

A SURVEY OF G AND K GIANTS AT THE SOUTH GALACTIC POLE

Chris Flynn

Mt. Stromlo and Siding Spring Observatories
Australian National University

John Bahcall

Institute for Advanced Study

K. C. Freeman

Mt. Stromlo and Siding Spring Observatories
Australian National University

ABSTRACT: We describe a sample of K giants extending to about 1.5 kpc at the South Galactic Pole. Important differences in the distribution of the stars in height Z and vertical velocity W are observed as a function of abundance and luminosity, which show that the stars, as selected, are not a simple isothermal tracer of the galactic potential. We examine the main features of the distribution of the sample in Z versus W, and discuss their implications.

1. INTRODUCTION

One of the classical problems in Astronomy is to determine the local density of matter, ρ_0, and the column density of matter in the disk near the Sun, Σ_0. The classical technique for obtaining these quantities *dynamically* is to measure the density falloff with height above the galactic plane Z in a "tracer" population of stars for which the distribution of the vertical velocities W at the plane is, hopefully, well understood. Recent advances in detector technology have meant that it is possible to obtain large samples of tracer stars at the galactic poles *in situ* extending to several kpc, and this is the focus of much of the work at this conference. The major advantage is that the velocity distribution and density of the tracer population as functions of height Z can be determined from the same sample.

In this talk, we will describe a sample of K giants extending to $Z \approx 1.5$ kpc in a region at the South Galactic Pole (SGP). The K giants offer several advantages. They are bright and very numerous, and their

A. G. Davis Philip and P. K. Lu (eds.)
The Gravitational Force Perpendicular to
the Galactic Plane   105 - 114
©   1989   L. Davis Press

radial velocities can be accurately measured. In addition, their abundances and luminosities can be accurately determined from DDO photometry.

The measurement of abundances and luminosities has been of great utility in understanding the sample. Important changes have been found in the distribution of the tracer stars in phase space (Z versus W) as a function of these quantities.

In section 3 we describe the sample and the observations we have made. Section 4 describes the classification of the stars using DDO photometry, and in section 5 we present some results showing the differences in the phase space distributions of the stars as functions of their luminosities and abundances. Some of the properties of the bulk of the stars and future work are discussed in section 6.

3. THE SAMPLE AND OBSERVATIONS

The candidate stars have been drawn from three sources, the Michigan Spectral catalog (Houk 1982, 1988), our own photographic photometry, and Eriksson's photoelectric catalog at the SGP (Eriksson 1978).

We chose the intermediate band DDO photometric system of McClure (1976) in order to remove dwarfs from the sample and to measure accurate abundances and luminosities for the giants. The four filters we used (41, 42, 45 and 48) are particularly suited to metal-rich giants with [Fe/H] $\geq$ -1.0. Below this abundance, the filters 38 and 35 are required in order to obtain accurate abundances and luminosities.

The bulk of the sample consists of HD stars from G 8 to K 5 brighter than luminosity class V in the Michigan Catalog (Houk 1982, 1988) in a 430 square degree region at the SGP[1]. There are just over 600 HD stars extending to V $\approx$ 10.2. For approximately 90 % of these, photoelectric DDO and BV photometry has been obtained with the 92", 40" and 24" telescopes at Siding Spring Observatory. Radial velocities, accurate to 3 km/s, have been obtained using the coude' focus of the 74" telescope at Mount Stromlo.

The other sources of data, our own BV photographic photometry and Eriksson's SGP catalog, cover about a quarter of the area of the HD sample and extend to V = 11.0. We selected objects with (B-V) > 0.9 (with appropriate allowance for color errors) as candidate giants. DDO and BV photometry and radial velocities have been obtained for $\approx$ 300 stars in these extended regions. These regions were primarily used to extend the survey to $\approx$ 1.5 kpc. In addition, they allowed us to determine that the HD catalog is complete to V = 9.2 in the color range 1.0 < (B-V) < 1.5.

---

[1]The classifications in Volume IV were kindly provided in advance of publication by Nancy Houk

4. STELLAR CLASSIFICATION

The DDO colors have allowed us to separate the giant and dwarf stars, particularly important for the objects in the extended regions, where no luminosity classifications are available.

We show in Fig. 1 the distribution of the program stars in the DDO colors C4245 versus C4548. These colors are *primarily* sensitive to effective temperature and luminosity respectively (see McClure 1976). The regions occupied by dwarfs and subgiants are marked by V and IV respectively, showing that the dwarfs can be easily removed. In the subgiant region, it is not possible to measure abundances and luminosities for all the program stars using the calibration of Janes (1975). The upper boundary of the subgiant region is defined by $M_v = 1.20$. All the "IV" and "V" objects have been removed.

The giants lie in the region marked III. Since most of the giants have been selected by spectral type and not by color, we have used our photoelectric BV photometry to determine that the sample is complete in the color range $1.0 < (B-V) < 1.5$, and these color limits have been marked on Fig. 1. Objects outside these color limits have been removed.

The stars marked with squares *may* be halo stars, with $[Fe/H] \lesssim -1.0$, although follow up DDO observations in the filters 35 and 38 are required to confirm this. These objects have been removed from the sample.

Stars marked by circles are "luminous giants" with $M_v < -0.6$ (as determined from DDO photometry). These have a substantially reduced velocity dispersion relative to the less luminous stars, and are discussed in the next section.

5. PHASE SPACE: ABUNDANCE AND LUMINOSITY EFFECTS

In this section, we show some important effects on the phase space distribution (Z versus W) of stars in the sample, as functions of their abundances and luminosities.

We note that the DDO absolute magnitudes have an associated error (from errors in the photometry) which is ≈ 0.5 for the brighter giants ($M_v < -0.6$) and ≈ 0.2 for the fainter giants ($1.2 > M_v > -0.6$). The abundances are accurate to ≈ 0.2 dex.

We plot in Fig. 2(a), for stars defined by $[Fe/H] > -0.5$ or "metal rich", the W velocity in km/s versus the absolute magnitude, M_v. Note the difference in the vertical velocity dispersion σ_w as a function of absolute magnitude. For 49 stars with $M_v < -0.6$, $\sigma_w = 14$

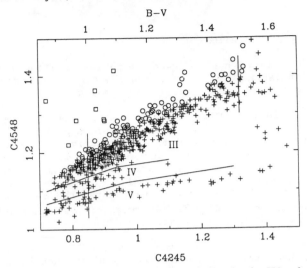

Fig. 1.   The distribution of the program stars in the DDO colors C4245 versus C4548.   Dwarfs fall in the region marked V.   The region marked IV contains   "subgiants", for which it is not always possible to measure abundance and luminosity.   The color completeness limits 1.0 < (B-V) < 1.5 are shown as two vertical lines (see text for discussion). The final sample giants lie between these color limits in the region marked III.   A few stars lie redward of (B-V) = 1.5; these are mostly later than K5 and have been removed, as the DDO system is not calibrated here.   The "luminous giants" discussed in the text are shown as circles, and *possible* halo stars are marked as squares.

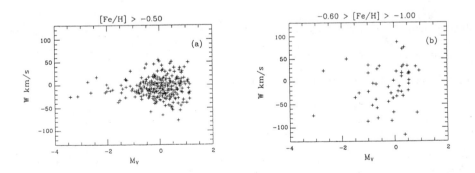

Fig. 2.   Plots of absolute magnitude, M_v versus the W velocity in km/s for the SGP sample.   In Fig. 2(a), we show the "metal-rich" stars. Note the difference in the vertical velocity dispersion σ_w as a function of absolute magnitude.   The velocity dispersion increases dramatically for the "metal-weak" stars in Fig. 2(b).

± 1 km/s while for 246 stars with $M_v > -0.2$, $\sigma_w = 19\pm1$ km/s.

In Fig. 2(b) we show W velocity versus M_v for the "metal weak" stars, with $-0.6 > [Fe/H] > -1.0$. Stars with abundances lower than $[Fe/H] = -1.0$ have been excluded as the four filters we employed do not give an accurate abundance below this limit. The velocity dispersion for 49 stars is $\sigma_w = 46$ km/s and is much higher than either subgroup in Fig. 2(a).

Support for this result at the SGP in Fig. 2(a) is shown in Fig. 3, in which we plot W velocity versus M_v for the "metal rich" stars ($[Fe/H] > -0.5$) in the Bright Star Catalog (BSC - Hoffleit 1982) with DDO photometry from McClure and Forrester (1981). The W velocities have been determined by projecting radial velocities and proper motions on to the Z direction, after calculating distances to each star from dereddened DDO colors. The expected error in W due to errors in the radial velocity and proper motions is approximately 10 km/s, about three times greater than the error for the SGP stars. (Note that the velocity dispersions we quote have been corrected for these errors.) The BSC stars strongly support our results at the SGP, (although the errors are larger due to the uncertainty of the BSC W velocities). For 147 stars with $M_v < -0.6$, $\sigma_w = 14\pm2$ km/s while for 450 stars with $M_v > -0.2$, $\sigma_w = 19 \pm1$ km/s.

How can we interpret these results? Firstly, the "metal weak" stars ($-0.6 > [Fe/H] > -1.0$) are in an abundance range associated in recent years with stars of the "thick disk" (see Freeman 1987 and references therein). We measure $\sigma_w = 46$ km/s for the objects - this dispersion is consistent with the observed scale height of the thick disk of approximately 1.5 kpc (Gilmore and Reid 1983). As discussed in Freeman (1987), the thick disk may be some kind of discrete relic of an early stage of the formation of the disk, either as an epoch of efficient star formation during the last stages of the collapse, or due to the capture of satellite galaxies. Alternatively, Norris (1987) has suggested that the thick disk may be better described as a hot, metal weak tail of the old disk.

As far as the measurement of the local density is concerned, it is more important to be sure these objects have had sufficient time to phase mix into the galactic potential. In fact, recent work indicates that the thick disk is as old as or a few Gyr older than the old disk (Norris and Green 1988, Carney et al. 1989). In either case, these objects should be phase mixed since they have oscillated in Z many times and therefore have a role as tracers of the potential.

We now turn our attention to the luminous stars with $\sigma_w \approx 13$ km/s in Figs. 2(a) and 3 (from the SGP and BSC respectively). It is possible that these are younger, higher mass stars than the bulk of the sample giants. If so, they will still be responding to the secular

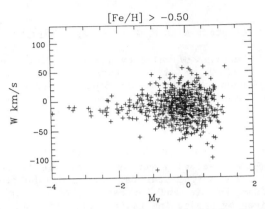

Fig. 3. Absolute magnitude, M_v versus W velocity for the "metal-rich" stars ([Fe/H] > -0.5) in the Bright Star Catalog, showing a similiar dependence on absolute magnitude as seen in our stars at the SGP. Note that we have removed stars with a likely error of more than 20 km/s in the W velocity due to the errors in the radial velocity or proper motion components.

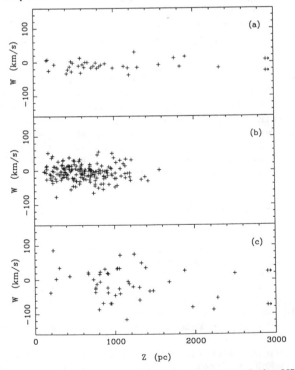

Fig. 4. Plots of Z versus W for three subgroups of the SGP sample. Fig. 4(a) shows the luminous metal-rich giants, (0.0 > [Fe/H] > -0.5 and M_v< -0.6). Fig. 4(b) shows the fainter metal rich giants, (0.0 > [Fe/H] > -0.5 and M_v > -0.2), and Fig. 4(c) shows the "metal-weak" giants (-0.6 > [Fe/H]   -1.0). These subgroups are discussed in the text.

mechanisms which produce the age velocity relation observed in the galactic disk (Fuchs and Wielen 1987). Of course, the secular heating is a continuous process, and it is likely that we are in fact sampling continuous change in Fig. 2(a).

We show in Fig. 4, the distributions in phase space (Z versus W) of each of the subsamples described above. Fig. 4(a) plots the luminous metal-rich giants, ([Fe/H] > -0.5 and M_v < -0.6), showing a cold central component extending to high Z in this magnitude limited sample. Fig. 4(b) plots the bulk of the sample, the fainter metal-rich giants, ([Fe/H] > -0.5 and M_v > -0.2), with a velocity dispersion we traditionally associate with old disk giants. Fig. 4(c) shows the metal weak giants (-0.6 > [Fe/H] > -1.0), with a velocity dispersion of approximately 46 km/s, and although this abundance range is not meant to be definitive, we associate the objects with the thick disk.

The distributions in Fig. 4 demonstrate the importance of obtaining abundances and luminosities for the giants. While it is true that the bulk of the sample exhibits the traditional properties of the old disk, we also see kinematically hotter "thick disk" and colder, possibly younger stars in the sample. Such properties are likely to feature in samples of F stars and K dwarfs, in addition to the K giants.

6. FUTURE WORK

The analysis of this sample depends on an accurately determined observational luminosity function for the sample giants, because there is a range in luminosity on the giant branch of several magnitudes. This is much more of a problem for the giants than for samples of K-dwarfs or even F-stars.

We have begun work on this problem by considering a reduced "core" sample of the tracer stars. As noted in section 3, the HD sample has been found to be magnitude complete to V = 9.2. These stars therefore belong to a well defined subset of the data, and provide a simple means to measure the density falloff. We use the following criteria to define the "core" sample : 0.0 < [Fe/H] < -0.5, V $\leq$ 9.2, and 1.2 > M_v > -0.2. These have been chosen to remove the "thick disk" stars and the luminous low-velocity dispersion stars.

We show in Fig. 5 the distribution of this "core" sample in phase space, Z versus W. The distribution is close to isothermal, with σ = 18 km/s, lending itself to the Bahcall (1984) technique for calculating, for various mass-models of the disk, the density gradient with Z of an isothermal population. In the case of a magnitude limited sample, the luminosity function of the "core" sample depends not only on the distribution of observed absolute magnitudes but also the density gradient of the tracer; it is for this reason that a near isothermal tracer simplifies the calculations. We are continuing work

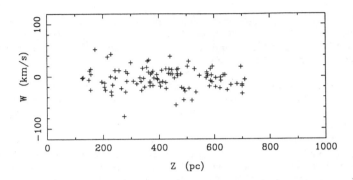

Fig. 5.   Plot of Z versus W for the "core" sample described in the text.

on this problem.

While this procedure will yield useful information about the bulk of the sample, it remains likely that the subgroups we have examined above are actually crude cuts in a continuous distribution. In this case, a self consistent solution for the mass profile of the disk and the luminosity function of the tracers will need to be derived from our measurement of their distribution in phase space, in order to measure the local and column densities of disk matter.

ACKNOWLEDGEMENTS

It is a pleasure to acknowledge discussions with Jesper Sommer-Larsen. CF is very grateful for financial assistance from the Institute for Advanced Study at Princeton, an Australian National University Postgraduate Scholarship, the Beckwith Trust and Western Connecticut State University.

REFERENCES

Bahcall, J. N. 1984 Astrophys. J. **287**, 926.
Carney, B. W., Latham, D. W. and Laird, J. B. 1989, Submitted to
    Astron. J..
Eriksson, P. -I. W. 1978 Uppsala Astr. Obs. Report **11**.
Freeman, K. C. 1987 Ann. Rev. Astron. Astrophys. **25**, 603.
Fuchs, B. and Wielen, R. 1987 In The Galaxy. Nato Advanced Study
    Institute, G. Gilmore and R. Carswell, eds., Reidel,
    Dordrecht.
Gilmore, G. and Reid, N. 1983 Monthly Not. Roy. Astron. Soc.
    202, 1025.
Hoffleit, D. 1982 Bright Star Catalogue, 4th Edition,
    Yale University Observatory, New Haven.
Houk, N. 1982, 1988 Michigan Spectral Catalog, Vols. III and IV.
    Braun-Brumfield, Ann Arbor.

Janes, K. A. 1975 Astrophys. J. Suppl. 29, 161.
McClure, R. D. 1976 Astron. J. 81, 182.
McClure, R. D. and Forrester, W. T. 1981 Pub. Dom. Astrophys. Obs.
   15, 439.
Norris, J. 1987 Astrophys. J. Letters 314, L 39.
Norris, J. and Green, E. M. 1988 Astrophys. J., in press.

DISCUSSION

RATNATUNGA: What are the errors in your absolute magnitude estimates as a function of M_v?

FLYNN: For luminous stars, with $M_v > -0.6$ the error is approximately 0.5 mag. For the stars with $M_v < -0.6$, the error is smaller, approximately 0.2 mag.

SCHECHTER: Can you isolate your "cold lump" by adjusting your metallicity limits?

FLYNN: No. The "cold" stars have a range in abundance from 0.3 to -0.5. They can only be cut away by using absolute magnitudes.

YOSS: Blanketing corrections can help establish the connection between M_v as a function of abundance.

LU: 1) How are your radial velocities determined? 2) How large an area does your survey cover?

FLYNN: The radial velocities are obtained on the 74 in coude' at Mgb with 6 km/s per channel. The survey covers 400 square degrees. The HD sample is complete to V = 9.2 In an approximately 100 square degrees the photographic survey goes to V = 11.0 mag.

WING: I'm concerned about the problem of determining absolute magnitudes that are independent of composition from DDO photometry. As you mentioned there is absorption by MgH in the 48 filter, and this is what separates dwarfs from giants. But what separates supergiants from giants is the CN (0,2) absorption in the 45 filter. So the DDO luminosity criterion depends upon both metallicity and O/C ratio, as well as luminosity itself. How do the DDO luminosity classes compare with the ones in the Michigan Spectral Catalogue? Were your "luminous giants" singled out there, too?

FLYNN: Yes, the class II giants were indeed bright in the DDO system, visually brighter than $M_v = -1.0$ The "luminous" giants we are removing from the sample go down to $M_v = -0.2$ and were mostly III's in the HD Catalog.

FLYNN: The broad depression feature found by Bond-Neff and studied by me may have some effect on the luminosity, since this depression is correlated to the color index (41-42).

GILMORE: I would like to remind you of the point raised by Fuchs and Wielen, that analyzing K_z data by fitting isothermal components is extremely sensitive to (immeasurably small) deviations from isothermality. Such deviations might be expected even if one did have perfect absolute magnitude and [Fe/H] data, due to (the well-established) dispersion in the age-metallicity relationship, even if for some reason the age-velocity dispersion relation were able to produce "isothermals". 2) Why have you gone to such a lot of trouble to throw away such a lot of your data, rather than analyze the whole observed distribution function? That seems a more efficient use of all your hard work.

FLYNN: Quite true - although Andrew Gould has shown us in his talk that the "thermal dispersion" of the isothermals can be estimated and corrected for. In fact, the components we have isolated in the sample demonstrate your point (especially the $M_v < -0.6$ stars). In answer to your second point, we have a problem to deal with in this sample - the broad luminosity function of the giants relative to other tracer samples such as F-stars or K-dwarfs. This complicates the measurement of the density falloff of the giants; in fact the luminosity function and the density falloff have to be derived self-consistently from the sample. Our isolation of a "core" sample with isothermal properties very much simplifies the calculation of the luminosity function, and will guide us to a solution using the whole sample.

NORDSTROM: Do you have an idea of the uncertainties in the velocities for the Bright Star Catalog?

FLYNN: We observed about 20 bright stars and found the error claim in the BSC of $\sigma_{RV} \approx 5$ km/s to be reasonable. However, there are over 800 stars in our BS sample - many radial velocities are dated and could be less accurate than claimed.

MAXIMUM LIKELIHOOD NUMBER OF DENSITY COMPONENTS IN THE GALAXY

Kavan U. Ratnatunga

Space Data and Computing Division
NASA Goddard Space Flight Center

1. INTRODUCTION

I summarize results from an analysis of the "Metallicity and Velocity Distribution of Giants Towards the Galactic Poles". This is indirectly important for K_z. For figures and a full analysis the reader is referred to a paper by Ratnatunga and Yoss (1989), which will be submitted to the Astronomical Journal.

John Bahcall requested me to show my bottom line first. In Fig. 1 I plot the metallicity, line-of-sight velocity distribution of a sample of 111 K-giants towards the galactic poles with z < 0.9 kpc and [Fe/H] > -0.5 . Although the distribution (looking at the kinematics only) is statistically isothermal (Bahcall 1984a) with a velocity dispersion of 20 km/s, the lower velocity dispersion of the more metal-rich giants may be a clue that reality may be hidden by small number statistics.

The 23 giants with [Fe/H] < 0 have a velocity dispersion σ_w = 11 ± 2 km/s in contrast to σ_w = 22 ± 2 km/s for the remaining 88 giants. When the sample is divided, using the distance from the galactic plane, the velocity dispersion of the lower and upper halves of this sample are 19 and 21 km/s respectively. With a rms error of 2 km/s in these estimates, the change is not significant. However the subsample giants with a lower velocity dispersion could be expected to have a smaller scale height and therefore if we measure the dispersion as a function of height we expect the population to show non-isothermality except for the small number statistics in this sample.

2. OBJECTIVE

The aim of the study is to use maximum likelihood rather than subjective judgment to determine the best representation of the observed metallicity, line-of-sight velocity distribution of a sample of giants towards the galactic poles. The study also aims to determine how many discrete isothermal components are required, or whether it is

A. G. Davis Philip and P. K. Lu (eds.)
The Gravitational Force Perpendicular to
the Galactic Plane   115 - 121
©   1989   L. Davis Press

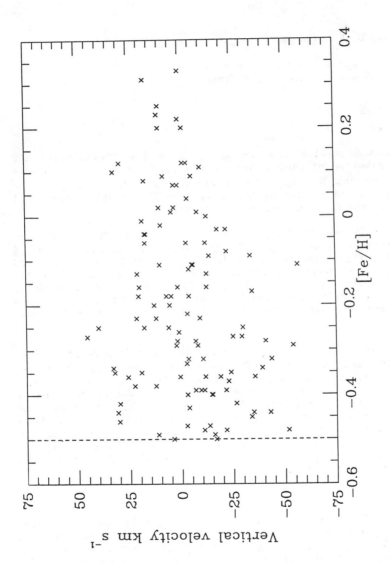

Fig. 1.   Sample of 111 NGP K giants used in Bahcall (1984b).

better represented by a continuous change in velocity dispersion with metallicity.

3. SAMPLE

The sample analyzed is a catalog of line-of-sight velocity, metallicity and distance for 441 spectroscopically selected giants near the galactic poles ($|b| > 80°$), taken mostly from Yoss, Neese and Hartkopf (1987).

The selected sample is incomplete in metallicity and distance from the galactic plane, both of which are estimated by a combination of many spectroscopic and photometric observable quantities. However, incompleteness is independent of the kinematics since proper motions were not used for selection of this sample. Line-of-sight velocity is an unbiased observable.

4. MODEL

The distribution function

$$f(\theta) = \Sigma_{i=1} \rho_i(z) \vartheta_i([Fe/H]) g_i(w)$$

is assumed to be a sum of $i = 1, \ldots, N$ discrete components, with separable distributions in metallicity $\vartheta_i([Fe/H])$, number density $\rho_i(\sigma)$, vertical velocity $g_i(w)$. Functions $g(w)$ and $\rho(z)$ are parameterized by the vertical velocity dispersion σ_w, which for each component is assumed independent of ($[Fe/H]$) and height (z).

The vertical density law for each component is defined dynamically using self consistent numerical solutions to the Boltzmann and Poisson equations in the presence of a dark halo (Bahcall 1984a). However due to conditioning for incompleteness the fit is insensitive to the exact form of density law assumed.

The line-of-sight velocity distribution is assumed to be Gaussian. The metallicity distribution of each component (as observed) is also assumed to be Gaussian over the [Fe/H] range that components mix, using Logit functions as developed by Nemec and Nemec (1989).

The fractional completeness at some ($[Fe/H],z$) on a star-by-star basis is assumed to be identical for each of the N discrete component and independent of the parameters of the model being maximized. I am in-fact assuming that the ($[Fe/H],z$) distribution of the sample without the observed line-of-sight velocities has no information about the distribution function.

5. LIKELIHOOD RATIO TEST

The maximum likelihood for a distribution fitted with n more parameters is necessarily larger since we have more degrees of freedom. Within some simplifying assumptions the likelihood ratio (difference in $\log_e(L)$) is a χ^2 distribution with n degrees of freedom.

The distribution of likelihood ratio can also be evaluated numerically using a lot of CPU and Monte Carlo catalogs. At NASA/GSFC/SDCD, I have access to two VAX 8820's associated with two currently launch-delayed satellites. I am able to analyze many simulated catalogs with exactly the same selection, incompleteness of the real data, but from a known distribution function.

For the Monte Carlo simulation I take each selection observable remove a random error to get a "true" value. Then using the set of "true" values I evaluate the expected line-of-sight velocity distribution and pick a random velocity from that expected distribution.

To compare Hypothesis A and Hypothesis B, generate Monte Carlo catalogs. 1) Make a simulation using Hypothesis A. Fit assuming A and B. 2) Make a simulation using Hypothesis B. Fit assuming A and B. Then look at the distribution of the likelihood ratio between the two fits and compare with the same ratio observed for real data. The distribution of likelihood ratio between fits assuming 3-components and 2-components on simulated catalogs with two discrete components follows the expected χ^2 distribution.

The distribution of likelihood ratio between fits assuming 2, 3, and 4 components on simulated catalogs with three discrete components matches the observed likelihood ratios fitted to the real sample. We infer that 3 discrete components are necessary and sufficient with over 95% statistical significance.

6. BEST FIT

The optimum representation has three discrete components: 1) A solar abundance thin disk with a σ_w = 12 $\pm$1 km/s; 2) A [Fe/H] ~ -0.20 $\pm$0.05 old disk with σ_w = 28 $\pm$2 km/s. 3) The metal-weak spheroid with σ_w = 54 $\pm$4 km/s. The local neighborhood fraction of the old disk giant number density is estimated to be roughly 16% $\pm$6%.

The next discrete component in the expansion of the velocity distribution has a velocity dispersion of σ_w = 49 $\pm$5 km/s and mean metallicity [Fe/H] = -0.66 $\pm$0.2. However this fourth component with typical "Thick disk" characteristics is not required to model the stellar content of this sample. It is not statistically separable from the metal-weakest component in this particular sample, which has been often used as confirmation for the existence of a thick disk.

7. CONTINUOUS CHANGE

The velocity dispersion function over metallicity is defined by the quadratic sum of velocity dispersions of components defined as in the discrete model.

$$\sigma^2([Fe/H]) = \Sigma_{i=1} \, \vartheta_i([Fe/H])\sigma_i^2.$$

The continuous function defined in this way has sufficient parameter freedom to fit the observed velocity dispersions. The distribution is not considered to be a mixture. Assuming that the function does not depend on height z, the solution is independent of distance to individual stars.

The distribution of likelihood ratio between using the continuous and discrete models is defined by fitting 100 Monte Carlo catalogs generated using the continuous and discrete distribution functions. The two distributions are well resolved showing that the ratio gives a reliable prediction of the nature of the distribution. The likelihood ratio observed for the real data is well beyond that observed for any of the simulated continuous catalogs.

8. MISSING DISK-LIKE-MASS

A similar likelihood ratio test using the Bahcall (1984b) missing disk-like mass proportionality constant P = 0 (no missing disk-mass) and P = 1 (50% missing disk-mass) shows that although the no missing mass is favored the likelihood ratio distributions overlap at the value observed in the real data. The missing disk-like mass fraction is not constrained by this sample.

9. DISCUSSION

The velocity dispersion and the mean metallicity of stars are both functions of age. Their dependences reflect the averaged history of chemical evolution in the Galaxy.

The reality of discrete components requires major events in the evolutionary history of our Galaxy such as 1) comparatively short bursts of active star formation; 2) accretion of satellite galaxies. The discrete nature of the catalog appears to indicate that the signature of such events seems to dominate the distribution over the continuous star formation which must also be present. Intrinsic large scatter in the age-metallicity-kinematic relationships appear to have not erased the signature of some major events of duration, short compared to the age of the Galaxy.

10. FUTURE PLANS

The procedure can be generalized to do maximum likelihood

analysis of any incomplete catalog of line-of-sight velocity and compare the various models for the distribution. For example 1) proper-motion selected samples. (Carney and Latham); 2) globular cluster system. Preliminary results for both have been derived, but not, the extensive tests required to report with confidence. I hope to also analyze Gilmore's K-dwarfs.

11. CONCLUSION

The distribution function for the Galaxy appears to be best represented by 3-discrete components rather than a continuous change of velocity dispersion with metallicity, if the latter function is assumed independent of height from the galactic plane.

K-giants with [Fe/H] > -0.5 are not a single isothermal. An old disk population with σ_w = 28 $\pm$2 km/s is isolated in addition to the young disk with σ_w = 12 $\pm$1 km/s in the sample of spectroscopically selected giants towards the poles.

REFERENCES

Bahcall, J. N. 1984a Astrophys. J. 287, 926.
Bahcall, J. N. 1984b Astrophys. J. 276, 156.
Nemec, A. F. L. and Nemec, J. 1989, in preparation.
Ratnatunga, K. U. and Yoss, K. M. 1989, preprint.
Upgren, A. R. 1962 Astron. J. 67, 37.
Yoss, K. M., Neese, C. L. and Hartkopf, W. I. 1987 Astron. J.
    94, 1600.

DISCUSSION

KING: If components have different distributions of absolute magnitudes, does your magnitude incompleteness not introduce a bias? In this connection, I am bothered by your assumption of equal a priori probabilities.

RATNATUNGA: The absolute magnitude incompleteness is correctly allowed for by the conditions on metallicity and distance. The equal probability assumption is a more subtle point. It is a standard procedure in most statistical studies of incomplete samples, and is the only assumption possible in the absence of any other information.

Flynn: Do you have absolutely no idea of the completeness of the sample at all?

RATNATUNGA: The sample of giants has been compiled from too many sources in the literature.

YOSS: You should be able to test King's question with your FAKE samples.

RATNATUNGA: Yes, I will try to formulate a suitable test to see the importance of the effect.

FREEMAN: In your 3-component fit, the two more metal-rich components each neatly straddle the boundaries of apparently visible components seen in Chris Flynn's data.   Your component one straddles the [Fe/H] > 0 component and the 0 > [Fe/H] > -0.5 component that Chris dismissed. Your components appear counter-intuitive: could you comment?

RATNATUNGA: It will be very interesting to subject Chris Flynn's full sample to this same analysis to see if it behaves differently.

NORRIS: What are the critical observational data required to permit your technique to make a definitive statement about the existence or otherwise of a thick disk component?

RATNATUNGA: More giants should be observed in the expected interface between the thick disk and spheroid, say -0.7 > [Fe/H] > -1.3 to see if better number statistics will be able to separate out the thick disk significantly.

GILMORE: I would recommend that you try your analysis on the sample analyzed by Wielen to determine the age-velocity dispersion relation. A recent discussion with the references is by Fuchs and Wielen (1987) in "The Galaxy", G. Gilmore and R. Carswell, eds.

RATNATUNGA: I agree I should try my analysis on as many samples as possible, including your K-dwarfs, which I hope you will send to me. It is only if many different catalogs lead to similar conclusions that we can start to accept the results.

KING: One must be very careful about using the original McCormick velocities.   In many cases the parallaxes fail to satisfy the Lutz-Kelker criterion that the error be less than 0.17 of the parallax. The distances of those stars are unknown and their space velocities are therefore worthless.   Fortunately Upgren is re-determining many of the parallaxes.

GOULD: Why do you think that using all of Bahcall's stars (plus some additional ones) you can't distinguish between a factor of two or so as Bahcall could?

RATNATUNGA: Bahcall used the sample of 111 K giants only to determine σ_w.   The density law was derived from Upgren's catalog of 4027 K-giants.   Incidentally, Strasbourg discovered last month, after a request I sent through Wayne Warren, that the Upgren (1962) K giant catalog had been accidentally deleted in 1978!   We are currently trying to obtain a copy of the original sent to Japan.

THE HALO POPULATION AS DERIVED FROM DDO PHOTOMETRY

Kenneth M. Yoss

University of Illinois

Carol L. Neese

Kitt Peak National Observatory, National
Optical Astronomy Observatories

David Bell

University of Illinois

1. OBSERVATIONS

DDO photometry has been obtained for an unbiased sample of 562 stars within 5° of the North Galactic Pole. The observations extend to V = 16. This subsample is drawn from the very large sample of Weistrop (1972, 1976, 1981). In addition to this faint-star set, we have observations of 580 SAO late-type stars, all of which are within 20° of the NGP. This subset is drawn from the published sample of Sandage and Fouts (1987), and from an unpublished sample by Griffin (1984). The parent samples consist of 877 G - K stars, for which radial velocities are available.

2. CLASSIFICATION

The DDO reduction and analysis of the data follows that described by Yoss, Neese, and Hartkopf (1987, hereafter YNH). [Eighty three of the 562 NGP stars were included in that paper.]

The statistics for the DDO classifications are given in Table I. In addition to the two samples discussed here, we include for information the numbers of stars observed at other cardinal points. The 364 stars listed under CENTER/ANTICENTER are discussed by Neese and Yoss (1988); the 445 stars in CYGNUS have DDO classifications and most have radial velocities (primarily from Sandage and Fouts 1987), and the results will be forthcoming. CCD spectra and DDO photometry for the 180 VELA stars were obtained in January, 1989 and currently are being reduced. The five entries in the table are: the total number in the sample, the number for which DDO classifications could not be made, the

A. G. Davis Philip and P. K. Lu (eds.)
The Gravitational Force Perpendicular to
the Galactic Plane   123 - 132
©   1989   L. Davis Press

TABLE I
DDO STATISTICS

SAMPLE	N (Obs)	N (No MK)	N (Dwarf)	N (Giant)	N (Pop. II)
Sandage/ Griffin	580	69	213 (42%)	285 (55%)	13 (3%)
NGP	562	98	303 (65%)	146 (32%)	15 (3%)
CYGNUS	445				
VELA	180				
CENTER/ ANTICENTER	364				

TABLE II
VELOCITY DISPERSION IN W

Sp. Range	[Fe/H]	Sandage/Griffin			NGP		
		N	<[Fe/H]>	σ_w	N	<[Fe/H]>	σ_w
G0 - G4	< -0.80	3	-1.51	65	1	-0.87	42
	-0.80 -0.45	2	-0.63	83	2	-0.68	11
	> -0.45	5	-0.07	25	15	-0.17	56
	all	10	-0.62	44	18	-0.26	52
G5 - K6	< -0.80	18	-1.39	48	7	-1.45	95
	-0.80 -0.45	39	-0.61	33	25	-0.64	47
	> -0.45	134	-0.13	19	46	-0.16	25
	all	191	-0.35	26	78	-0.43	42
G0 - K6	< -0.80	21	-1.41	49	8	-1.38	90
	-0.80 -0.45	41	-0.61	35	27	-0.65	46
	> -0.45	139	-0.13	19	61	-0.16	35
	all	201	-0.36	27	96	-0.40	44

number of dwarfs, the number of giants, and the number of metal-poor giants, which are referred to here as Population II stars. This last category is loosely defined as giants for which [Fe/H] is derived through the Δ4548' procedure (Janes 1979), the values of which generally are < -1.00. Below the numbers of the last three categories (and in parentheses), the percentages of the classified samples are given.

The DDO colors of the stars not classified fall beyond the classification grids (Janes 1975); however, in many instances consistent multiple observations are available, and in general the (B-V) colors are consistent with G - K types. Spectroscopic classifications of those stars will be made, in an effort to determine their true nature.

It is not surprising that the NGP sample includes a larger fraction of dwarfs (65%) than does the Sandage/Griffin sample (42%), since it extends to faint stars, while the Sandage/Griffin sample is of bright stars (V < 10; z < 1 kpc). Both samples contain 3% Pop. II giants, but relative just to the sample of giants and dwarfs alone (minus the unclassified), the percentages are 4% and 9% for Sandage/Griffin and NGP respectively.

3. ABUNDANCES

The relationship between z and [Fe/H] for the giants of the NGP sample is shown in Fig. 1; this is similar to Fig. 1a of YNH, except that the G 0 - G 4 giants also are included, a large fraction of which fall within the upper left portion of the diagram, delineated by the dashed lines, and referred to as the "zone". The significance of this region is discussed in YNH in greater detail, but in general it is a region which few stars habitate (with the exception of the G 0 - G 4 group). The exclusion of the G 0 - G 4 giants in YNH Fig. 1 was because the DDO classifications were uncertain. However, as discussed below, the spectral types and luminosities (and thus distances) have been confirmed spectroscopically. The abundances still are somewhat uncertain.

4. W VELOCITY DISPERSIONS

The velocity dispersions of the giants with available W velocities are summarized in Table II, as functions of [Fe/H]. Previously the stars were grouped traditionally into "components", such as "halo", "thick disk", etc, with the abundance boundaries chosen to a great extent on the basis of previous work. In the present instance, both the number of components and the boundaries are chosen through a maximum-likelihood procedure, which for this particular NGP sample specifies the [Fe/H] ranges for only three populations (Ratnatunga 1989). In the table, the stars are grouped into two spectral ranges (G 0 - G 4 and G 5 - K 6) and the total (G 0 - K 6). The results are further grouped according to the three abundance ranges determined

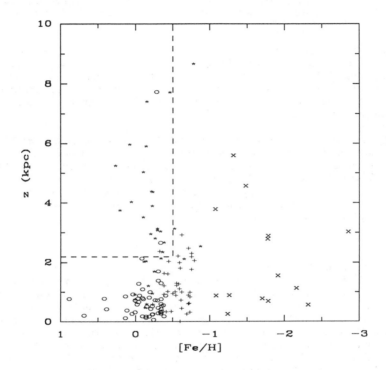

Fig. 1.   [Fe/H] versus z for the NGP giants. The symbols are:
* = G 0 - G 4 giants; o = G 5 - K 6 giants; + = stars with [Fe/H]
= mean of the δCN and the Δ4548' calibrations; x = stars
with [Fe/H] based on Δ4548' alone. The upper left region boxed
with the dashed lines is described in YNH.

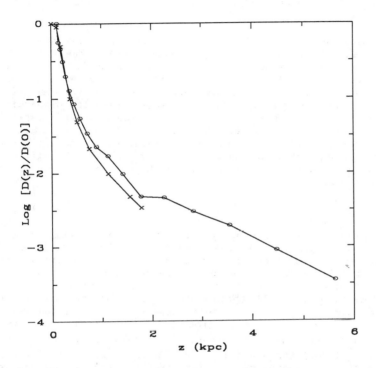

Fig. 2. D(z) versus z for the NGP Pop I G 5 - K 6 giants. The
symbols are: o = NGP; x = Elvius.

through the maximum-likelihood procedure. The number in each subsample, <[Fe/H]>, and σ_w are given for both Sandage/Griffin and the NGP. In several cases, the number is too small for a meaningful velocity dispersion, and mean values for W are then given. Unfortunately the sample sizes are not sufficient to divide them into distance groups (the Sandage/Griffin sample does not extend beyond 1 kpc in any case).

The changes in velocity dispersions with abundance are similar to those changes exhibited in other samples (e.g. YNH). The G 0 - G 4 group is of particular interest, as it exhibits a large velocity dispersion (YNH, Yoss 1988). This is marginally evident for both Sandage/Griffin and the NGP when all abundances are combined (last entries of each spectral range), with the largest velocity dispersions in the G 0 - G 4 group. Many of those stars also exhibit solar-type [Fe/H] and large z values (see Fig. 1, the "zone" population). Spectroscopic classifications confirm the spectral and luminosity classes (Yoss 1988), but confirmation of the abundances is difficult with low-resolution spectra. Detailed abundance analyses should be made on these stars.

5. D(z) and $K_z(z)$

A preliminary stellar density (number of stars/cubic parsec) toward the galactic pole, D(z), has been derived for the NGP Pop. I giants. Incompleteness factors were derived by comparison of the Weistrop sample and the total observed sample as a function of V; those incompleteness factors then were applied to the observed G 5 - K 6 giants (luminosity classes II through IV). The adjusted numbers of stars per 1/2 mag interval, A(m), were compared to calculated values through the (m - log π) process (Bok 1937); the differences were minimized by adjusting D(z). The adopted luminosity function $\phi(M_v)$ was that of McCuskey (1967) for G 8 III - K 3 III, $M_v = +1.0$, $\sigma = \pm0.6$ mag. This is clearly an approximation for the present giant sample. However, the sample is sufficiently small that we have included a large range in luminosities (luminosity classes II through IV), and until the sample can be increased substantially for now it is probably as good an approximation as is available. The resulting D(z) is shown in Fig. 2, with that of Elvius (1965) for comparison. The sharp dip and leveling off centered at z = 2 kpc cannot be forced out by altering the density at those distance without increasing the discrepancy between the observed and calculated A(m).

The galactic restoring force acting upon a star at distance z above the galactic plane can be expressed by the Boltzmann equation:

$$K_z(z) = \frac{1}{D(z)} \frac{d}{dz} [\sigma_w^2 \ D(z)], \qquad (1)$$

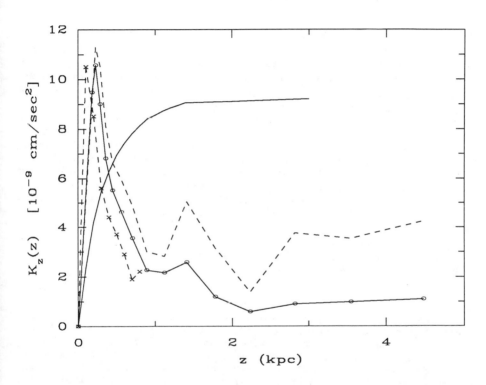

Fig. 3. $K_z(z)$ versus z. The symbols are: --- = NGP, eq. 1;
o = NGP, eq. 2, x = Upgren; _____ = Oort.

where $\sigma_w = f(z)$.

Alternately, assuming a dynamical steady state, σ_w is not a function of z, and:

$$K_z(z) = \frac{\sigma_w^2}{\text{mod}} \frac{d}{dz} [\log D(z)], \qquad (2)$$

where $\text{mod} = \log_{10} e = 0.43429$.

$K_z(z)$ has been determined for both cases. For case (1), $\sigma_w = 10 + 10z$ km/s, to $z = 2$, then $\sigma_w = 40$ km/s for $z > 2$, where z is in kpc. This relationship is consistent with the variation of σ_w found by YNH (Fig. 5). The result is shown in Fig. 3, along with the standard Oort (1965) relationship. To make a more direct comparison with previous solutions, we also show $K_z(z)$ based on equation (2), where $\sigma = 20$ km/s, a value used by both Upgren (1962) and Bahcall (1984). This second result, as well as the Upgren solution, also are shown in Fig. 3. Our results are similar to that of Upgren, and easily can be brought into better agreement by adjusting $\phi(M_v)$ which was rather arbitrarily chosen for this very preliminary and illustrative solution. Both are in sharp disagreement with the standard Oort function, even within the z < 1 kpc range, which is critical for the solution to the local stellar density $\rho(0)$. We should point out that a series of $K_z(z)$ solutions are available (e. g. Perry 1969), and none are in agreement with the Oort function. (See cover picture.)

6. FUTURE WORK

The need for a larger sample is evident, both in terms of DDO photometry and radial velocities. A CCD/grism camera is now available at Mount Laguna Observatory, with resolutions of 4.2 and 7.2 Å. Spectra of 16th mag stars are obtained in about 10 minutes, with acceptable S/N. Both DDO and (B-V) colors are extracted from the spectra with standard passbands, with results equal to standard filter photometry for the brighter stars, and with superior (and faster) results for stars fainter than 15th mag. With this technique, we will increase our faint-star sample substantially during the next observing season.

The CCD also is used with our Cassegrain spectrograph; acceptable spectra to B = 14.5 mag are obtained in 30 minutes with a resolution of 0.6 Å. Radial velocities as well as spectroscopic classifications thus will be added to our sample.

Particular attention must be paid to the intriguing group of G 0 - G 4 giants with "solar-like" abundances. DDO photometry is not ideal for this spectral range, and our future effort for this group includes four-color photometry, and eventually high-resolution spectra for detailed abundance solutions.

REFERENCES

Bahcall, J. N. 1984 Astrophys. J. 287, 926.
Bok, B. J. 1937 The Distribution of Stars in Space, University of Chicago Press, Chicago.
Elvius, T. 1965 in Galactic Structure, A. Blaauw and M. Schmidt, eds., University of Chicago, Chicago, p. 41.
Griffin, R. F. 1984, private communication.
Janes, K. A. 1975 Astrophys. J. Suppl. 29, 161.
Janes, K. A. 1979 in Problems of Calibration of Multicolor Systems, A. G. D. Philip, ed., Dudley Obs. Rep. No. 14, p. 103.
McCuskey, S. W. 1967 Astron. J. 72, 1199.
Neese, C. L. and Yoss, K. M. 1988 Astron. J. 95, 463.
Oort, J. H. 1965 in Galactic Structure, A. Blaauw and M. Schmidt, eds., University of Chicago, Chicago, p. 455.
Perry, C. L. 1969 Astron. J. 74, 139.
Ratnatunga, K. U. 1989 in The Gravitational Force Perpendicular to The Galactic Plane, A. G. D. Philip and P, K. Lu, eds., L. Davis Press, p. 115.
Sandage, A. R. and Fouts, G. 1987 Astron. J. 93, 592.
Upgren, A. R. 1962 Astron. J. 67, 37.
Weistrop, D. 1972 Astron. J. 77, 366.
Weistrop, D. 1976 Astrophys. J. 204, 113.
Weistrop, D. 1981, Private communication.
Yoss, K. M., Neese, C. L. and Hartkopf, W. I. 1987 Astron. J. 94, 1600.
Yoss, K. M. 1988 20th General Assembly, International Astronomical Union.

DISCUSSION

UPGREN: Ken, you had an early viewgraph that went by that was more like a railroad map of the state of Ohio. Are you going to end up laying another layer of track on that?

YOSS: Oh, that one (K_z vs z for previous solutions). It was marvelous, wasn't it? I really do not care where our relation goes, which incidently agrees well with yours. Gilmore's relation is close to that of Oort.

NORRIS: Is that going to be on the cover of the book?

NORRIS: Could you tell us what the color range was of your G0 stars?

YOSS: The colors are more consistent with the colors of G0 dwarfs than with G0 giants. We are going to do four-color photometry on these stars, which will give better classifications. We still would like to get Kitt Peak time to derive abundances. Our radial velocities are certainly all right.

NORRIS: Can you rule out that they are real horizontal-branch stars such as you see in M 3?

YOSS: I can not be sure at this time. The spectra say that they are G0, on the classic MK classification system.

RATNATUNGA: If you have spectra, isn't it better to look at the spectra and determine abundances rather than get DDO colors from the spectra?

YOSS: With our spectra we did not derive DDO colors but used line ratios and abundance indices (such as CN anomaly and overall strength of the metal lines). With our grism spectra we did derive DDO colors, quite successfully. We would like to do four-color this way but the grism spectra do not go far enough into the blue. We are going to try four-color CCD with standard filters, particularly for these G0 stars. I understand from Dave Philip that it really works with high accuracy. Next year we will do four-color with filters and DDO with the grism.

ATTACKING THE K_z PROBLEM WITH STÄCKEL POTENTIALS

Thomas S. Statler

JILA, Univ. of Colorado

I. INTRODUCTION

As Gilmore and Kuijken (1989) have emphasized, an important ingredient in the analysis of the dynamics of disk stars more than ~ 1 kpc above the midplane is that quantity which we are most accustomed to seeing as the cross term in the Jeans equation:

$$\rho \frac{\partial \phi}{\partial z} = \frac{\partial}{\partial z}(\rho \sigma_{zz}^2) + \frac{1}{\varpi} \frac{\partial}{\partial \varpi}(\varpi \rho \sigma_{\varpi z}^2). \tag{1}$$

The factor $\sigma_{\varpi z}^2$ in the last term on the right-hand side of equation (1) is the cross term, which is related to the tilt and the shape of the velocity ellipsoid. Current observations tell us very little about these properties of the velocity distribution outside the solar neighborhood, but we expect that the ellipsoid tilts in a systematic way with height z above the plane, somewhere between the following two extremes: (1) the ellipsoid remains parallel to the disk; (2) the ellipsoid points at the galactic center. Case (1) implies that the potential ϕ separates in cylindrical coordinates ϖ, θ, z) according to $\phi = \phi_\varpi(\varpi) + \phi_z(z)$. In such a potential the Oort integral, or z energy, is an exact isolating integral for all orbits, and the cross term is identically zero. The extent to which the cross term is non-zero indicates the extent to which the z energy fails as a third integral. Case (2) implies that the potential is spherical, and the total angular momentum L^2 is an isolating integral. Each of these limits, however, is unnecessarily restrictive. The general case, in which the velocity ellipsoid points neither parallel to the disk nor exactly at the galactic center, can be treated without sacrificing analytic tractability by assuming that the potential separates in spheroidal coordinates, i.e. is a Stäckel potential.

2. STÄCKEL POTENTIALS

2.1 Prolate Spheroidal Coordinates

A. G. Davis Philip and P. K. Lu (eds.)
The Gravitational Force Perpendicular to
the Galactic Plane   133 - 142
©   1989   L. Davis Press

Fig. 1a shows typical coordinate surfaces in the meridional plane for the spheroidal system. Surfaces of constant λ are prolate spheroids elongated along the symmetry axis; surfaces of constant ν are hyperboloids of two sheets everywhere orthogonal to the spheroids. All coordinate surfaces share the foci located at $z = \pm z_0 = (\gamma - \alpha)^{1/2}$, where α and γ are parameters. Without loss of generality one can arbitrarily set $\alpha = -1$ kpc^2, which simply establishes the length scale. That choice having been made, the location of the foci then determines γ, or vice versa.

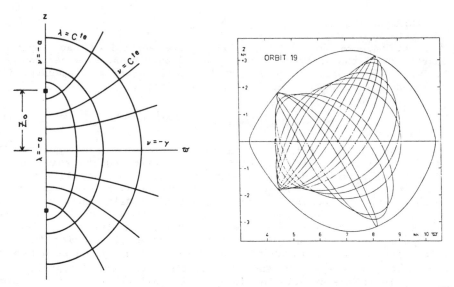

Fig. 1.   (a) Spheroidal coordinates in the meridional plane. All coordinate surfaces are symmetric about the z axis.   (b) Ollongren's "Orbit 19" in the meridional plane of the Schmidt potential. Note the similarity between the orbit envelope and the confocal surfaces of frame (a).

The origin is given by $\lambda = -\alpha$, $\nu = -\gamma$, the equatorial plane by $\nu = -\gamma$, and the z axis by $\lambda = -\alpha$ between the foci and $\nu = -\alpha$ outside. Notice that consequently λ and ν satisfy the inequality $-\gamma \leq \nu \leq -\alpha \leq \lambda$. Near the origin the coordinates are approximately cylindrical, and at large radii they are nearly spherical. The transition happens in the vicinity of the foci; thus in the limit $z_0 \to \infty$ the coordinate system is cylindrical everywhere, and in the limit $z_0 \to 0$ it is spherical everywhere.

2.2 Potentials and Integrals

In the prolate spheroidal coordinates (λ, θ, ν), an oblate

Stäckel potential has the form:

$$\phi = \frac{F(\lambda) - F(\nu)}{\lambda - \nu}, \tag{2}$$

where $F(\tau)$, for $\tau = \lambda$ or ν, is (for the moment) an arbitrary function. Actually, since λ and ν occupy disjoint intervals of the real line, $F(\tau)$ is really two such functions. One has a certain amount of freedom in specifying $F(\tau)$; for instance, if one agrees to define $F(\nu)$ so that, in the limit $\nu + \gamma \to 0$, $F(\nu) \propto \nu + \gamma$, then $F(\lambda)$ is determined completely by the circular velocity in the equatorial plane and a constant of integration:

$$V_c^2(\varpi) = -2(\lambda + \alpha)\frac{d}{d\lambda}\left(\frac{F(\lambda)}{\lambda + \gamma}\right), \qquad \varpi^2 = \lambda + \alpha. \tag{3}$$

Thus all constraints on the potential set by the rotation curve are automatically satisfied once the choice of $F(\lambda)$ is made. Then the function $F(\nu)$ can be chosen independently to produce the desired disk-halo structure, with one important caveat: one generally will not be able to produce a globally good representation of the galactic potential, but instead must confine oneself to a region well away from the foci. Since most disk stars are on orbits of low eccentricity, this is not at all a serious restriction.

The essence of the Stäckel potentials is that all orbits are bounded by coordinate surfaces (de Zeeuw 1985). The relevance of this fact to galactic dynamics can be seen in Fig. 1b, which shows Ollongren's (1965) "Orbit 19" calculated in the Schmidt (1956) potential. The similarity of the orbit envelope to segments of the confocal surfaces in Fig. 1a suggests that one may be able to obtain quite an accurate representation of the galactic potential near the solar circle with a potential of Stäckel form.

All orbits in a Stäckel potential have three exact isolating integrals; in terms of the cylindrical coordinates (a, θ, z) and the corresponding velocities (Π, Θ, Z), they are:

Energy : $\quad E = \phi - \frac{1}{2}(\Pi^2 + \Theta^2 + Z^2),$

Angular momentum : $\quad I_2 = \frac{1}{2}R_0^2\Theta^2,$ $\tag{4}$

Third integral : $\quad I_3 = \psi - \frac{1}{2}\frac{z^2}{z_0^2}(\Pi^2 + \Theta^2) - \frac{1}{2}\left(\frac{\varpi^2}{z_0^2} + 1\right)Z^2 + \frac{\varpi z}{z_0^2}\Pi Z,$

where ψ is a function, related to the potential, that behaves like ϕ_z in the cylindrical case. In fact, in the limit $z_0 \to \infty$, ψ becomes exactly ϕ_z and I_3 turns into the Oort third integral.

2.3 Parametrized Galactic Potentials

136   Thomas S. Statler

To apply Stäckel potentials to the K_z problem, one would like to have a set of such potentials parametrized by, say, the focal position z_0 (i.e. the velocity ellipsoid orientation), and the total local midplane density ρ_0 and surface density Σ_0. Then the objective of any modeling procedure is to find the values of those parameters most consistent with the kinematics of some observed sample of tracer stars toward a galactic pole.

The function $F(\lambda)$ that corresponds to an exactly flat rotation curve is:

$$F(\lambda) = -\frac{V_\infty^2}{2}(\lambda + \gamma)\ln\frac{\lambda + \alpha}{R_0^2}. \tag{5a}$$

However, it is counterproductive to try to match the rotation curve between the foci, since the fit to the true potential will be bad in that region regardless, and such an attempt tends to make matters worse near the Sun. A helpful remedy is to use the rotation curve:

$$V_c(\varpi) = V_\infty[1 + (z_0/\varpi)^4]^{-1/4},$$

generating

$$F(\lambda) = -\frac{V_c^2}{2}(\lambda + \gamma)\ln\frac{\lambda + \alpha + [(\lambda + \alpha)^2 + z_0^4]^{1/2}}{R_0^2 + [R_0^4 + z_0^4]^{1/2}}. \tag{5b}$$

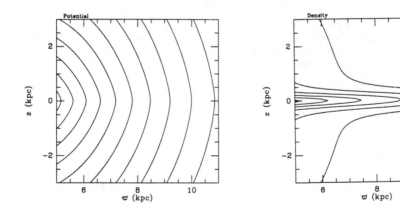

Fig. 2. A model galactic potential of Stäckel form. (a) Equipotentials near the solar circle. (b) The density that generates the potential in (a). Contours are separated by a factor of two in density.

As for $F(\nu)$, a choice that mimics the potential used by Kuijken and Gilmore (1989b), in which K_z is linear in z at both small and large z (but with different slopes) is

$$F(\nu) = -\alpha V_c^2 \left(-\frac{z_0^2}{\nu + \alpha} \right) \left[\left(\frac{S^2}{C^2} + S\frac{\nu + \gamma}{z_0^2} \right)^{1/2} - \frac{S}{C} \right], \tag{6a}$$

where C and S are related, respectively, to the central density ρ_0 and the surface density Σ_0. However, the requirement that $K_z \propto z$ at large z implies that the halo is unrealistically cylindrical. A better (and simpler!) function is

$$F(\nu) = -\alpha V_c^2 \left[\left(\frac{S^2}{C^2} + S\frac{\nu + \gamma}{z_0^2} \right)^{1/2} - \frac{S}{C} \right]. \tag{6b}$$

The potentials given by equations (5a) and (6a) are used in the models presented in section 4 and are described further in Statler (1989). The improved potentials using equations (5b) and (6b) will be described elsewhere (Statler 1990). Fig. 2 shows a potential of the latter type with $(z_0, C, S) = (0.2$ kpc, $500, 200)$ and the corresponding total density.

3. DISTRIBUTION FUNCTIONS

3.1 Generalized Isothermal Components

Since velocity distributions for stars in the solar neighborhood - at least for the well-mixed species - seem to be approximately ellipsoidal and Gaussian, it is natural to ask for the simplest function of the integrals (E, I_2, I_3) consistent with a velocity distribution at the solar position of the form

$$g(\Pi, \Theta, Z) \propto \exp\left\{ -\frac{1}{2\sigma_\varpi^2}\Pi^2 - \frac{1}{2\sigma_\theta^2}(\Theta - \Theta_m)^2 - \frac{1}{2\sigma_z^2}Z^2 \right\}. \tag{7}$$

The answer is:

$$f(E, I_2, I_3) = \exp\left[\frac{E}{\sigma_\varpi^2} + \frac{1}{R_0^2}\left(\frac{1}{\sigma_\varpi^2} - \frac{1}{\sigma_\theta^2} \right) I_2 + \frac{\Theta_m}{R_0\sigma_\theta^2}\sqrt{2I_2} \right. $$
$$\left. + \frac{1}{(R_0/z_0)^2 + 1}\left(\frac{1}{\sigma_z^2} - \frac{1}{\sigma_\varpi^2} \right) I_3 \right]. \tag{8}$$

Above the equatorial plane, the velocity distribution in such a "generalized isothermal" is a tilted ellipsoid:

$$f = B \exp\left\{ -h^2\Pi^2 - k^2[\Theta - V_r]^2 - l^2 Z^2 + n\Pi Z \right\}, \tag{9}$$

where B, h^2, k^2, l^2, n, and V_r are all algebraic functions of (ϖ, z). That these distributions can be written explicitly in terms of position and velocity means that dynamical models can be constructed that are exact solutions of the Boltzmann equation and that incorporate consistently both the tilt and the change of shape of the velocity ellipsoid. In the models discussed in section 4, distribution functions for tracer stars are found as linear combinations (with possibly negative coefficients) of a family of generalized isothermal components having fixed values of σ_θ and θ_m representative of the solar neighborhood and variable values of σ_ϖ and σ_z.

3.2 Building Basic Disks

A difficulty with the generalized isothermal components is that the density distribution of one such component in the model potentials is not a disk, but rather a flared torus. That this should be the case is obvious: first, the Z dispersion in the midplane declines only slowly with radius, while the disk restoring force drops exponentially; second, asymmetric drift, a hallmark of a differentially rotating disk, is absent in a generalized isothermal because the Π and θ distributions are independent. As a result, the radial gradient of the density, which contributes to the cross term in the Jeans equation, tends to be overestimated. A desirable refinement is to replace the generalized isothermals as the "basic building blocks" of the models with combinations of components that more resemble disks of various sorts (e.g. flaring, tapering, thick, thin, etc.).

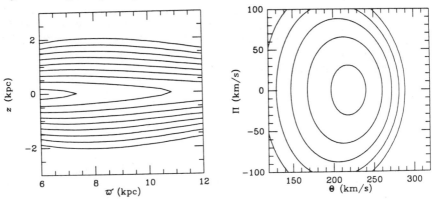

Fig. 3. An example of a "basic disk", made from a one-dimensional continuum of generalized isothermal components. (a) Density in the meridional plane. (b) Bottlinger diagram (the Π, θ velocity distribution) for this disk. In both frames, contours are separated by a factor of e.

An example of such a "basic disk" is shown in Fig. 3a. The distribution function is a one-dimensional path integral through the space of parameters (σ_ϖ, σ_θ, σ_z, θ_m); details of the construction of basic disks and their use in models will be published elsewhere (Statler 1990). This disk has an exponential scale length in the equatorial plane of 3.5 kpc, and a scale height of about 400 pc, very nearly independent of radius. All local velocity dispersions decline exponentially with radius, in qualitative agreement with the observations of Lewis and Freeman (1989). Fig. 3b shows a Bottlinger diagram for this disk; note that high-velocity stars tend to lag behind the mean rotation, as observed in the Galaxy.

4. SIMPLE MODELS AND THE ERROR BUDGET ON Σ_0

In order to test the modeling procedure and estimate the mean error on estimates of Σ_0 derived from newly available data (Kuijken and Gilmore 1989b), models were made for an artificial data set generated from a known potential and distribution function (Statler 1989). Model distribution functions were calculated, from the density distribution of stars in the data set, as linear combinations of generalized isothermal components taken from different one-parameter libraries. Each library was chosen so as to impose different trends on the shape of the velocity ellipsoid above the midplane. Adapting the procedure of Kuijken and Gilmore (1989a), the models were gauged according to the likelihood of the "observed" set of velocities in the model distributions.

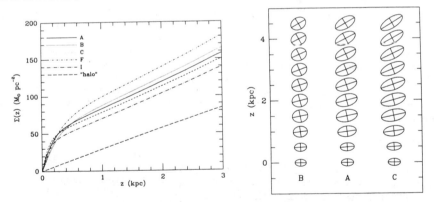

Fig. 4. Some properties of the models discussed in section IV. (a) Integrated column density between ±z as a function of z. The disk surface density for each model is the difference between the model curve and the curve marked "halo". (b) Behavior of the velocity ellipsoid above the equatorial plane for three of the models shown in (a).

Fig. 4a shows the total integrated density between ±z as a

function of z for five of the models. The curve marked "A" indicates the best-fit model using the library that correctly matches the "true" distribution for the artificial data; this model recovers the "true" galactic potential. The curves marked "B" and "C" represent the best-fit models obtained from incorrect libraries, which cannot be ruled out on the basis of their likelihood values (i.e. they are nearly as good fits as model A). The shape and orientation of the velocity ellipsoids for models A, B and C are shown in Fig. 4b. The curve marked "halo" is the column density of the halo alone; thus the disk surface density in each case is given by the asymptotic difference between the model curve and the halo curve.

Shot-noise errors were estimated from Monte Carlo simulations using different random samples of 500 stars drawn from the same parent distribution and potential. It was found that such errors contribute about a 21% uncertainty to the derived value of Σ_o. This is the largest single contribution to the total uncertainty; to halve it would require a sample of tracer stars three times larger. The unknown behavior of the velocity ellipsoid (the differences between models A, B, and C) adds another 10% in quadrature. (The effect is systematic, but we have no handles on either the sign or the magnitude.) Kuijken and Gilmore (1989b) estimate that uncertainties in the rotation curve and in the value of R_o contribute an additional 16%. Further complications arise from non-equilibrium effects such as scattering from spiral arms, the warp of the disk, the opening of the local superbubble in the ISM, and the possibility of large resonant orbit families, each of which can contribute at the 10% level. Consequently, the mean error on the current best estimates of Σ_o must be at least $\pm 30\%$. This conclusion agrees fairly well with that of Gould (1989a,b).

The large uncertainty implies that the presence or absence of substantial amounts of disk dark matter can still not be ruled out at better than the 1.5 σ level. To remedy the situation, we will need both larger tracer star samples to reduce the shot noise error and observations on lines of sight with $|b| < 60°$ to constrain the velocity ellipsoid.

REFERENCES

de Zeeuw, T. 1985 Monthly Not. Roy. Astron. Soc. **216**, 273.
Gilmore, G. and Kuijken, K. 1989 in The Gravitational Force Perpendicular to the Galactic Plane, A. G. D. Philip and P. K. Lu, eds., L. Davis Press, Schenectady, p. 61.
Gould, A. 1989a in The Gravitational Force Perpendicular to the Galactic Plane, A. G. D. Philip and P. K. Lu, eds., L. Davis Press, Schenectady, p. 19.
Gould, A. 1989b Astrophys. J. **341**, 748.
Kuijken, K. and Gilmore, G. 1989a Monthly Not. Roy. Astron. Soc., in press.
Kuijken, K. and Gilmore, G. 1989b Monthly Not. Roy. Astron. Soc.,

in press.
Lewis, J. R. and Freeman, K. C. 1989 Astron. J. **97**, 139.
Ollongren, A. 1965 in Stars and Stellar Systems, Vol. 5, Galactic
    Structure, A. Blaauw and M. Schmidt, eds., University of Chicago
    Press, Chicago, p. 501.
Schmidt, M. 1956 Bull. Astron. Inst. Netherlands **13**, 15.
Statler, T. S. 1989 Astophys. J. **344**, in press.
Statler, T. S. 1990, in preparation.

DISCUSSION

KING: I got lost at the very end where you managed to satisfy the Lewis-Freeman effect. Did you give up your isothermal distribution?

STATLER: Yes. Lewis and Freeman's observation is that $\langle \Pi^2 \rangle$ decreases exponentially with ϖ, from which they infer that $\langle Z^2 \rangle$ probably does something similar. So we already know the disk can't be strictly isothermal. The velocity distribution in the model disk I showed has approximately-Gaussian tails, but is neither isothermal nor exactly ellipsoidal, as you can see from the Bottlinger diagram.

SCHECHTER: Does the fact that Ollongren's Orbit 19 has an inner edge which is not part of an ellipse give us a handle on the limitations of Stäckel potentials?

STATLER: Yes, but it's a complicated handle. The extent to which the orbit envelope is not formed by confocal conics mainly reflects differences in the derivatives of the potential near the turning points. The shape of the inner and outer (as opposed to top and bottom) edges is most sensitive to the variation of the radial force with z, and so that particular attribute of the Schmidt potential would not necessarily prevent one from using a Stäckel potential on the vertical dynamics.

FREEMAN: Say you have a "realistic" galactic potential which you approximate by a Stäckel potential. The Stäckel potential defines its own coordinate surfaces. Now say you run orbits of different E, J in the "realistic" potential; these orbits define a set of "effective" coordinate surfaces. Can you comment on how well the two sets of surfaces (Stäckel and orbit-defined) compare: e.g., how well would inferences about $\sigma_{\varpi z}$ ϖ correspond?

STATLER: The short answer is, give me a galactic potential and I'll tell you. What you suggest is probably the best way to measure the difference between some non-separable potential and a Stäckel fit, but offhand I couldn't say exactly what the relation between velocity dispersion and envelope position is. The Stäckel fit will be better the smaller a region one is interested in, and will also tend to be better for rounder potentials than for flatter ones. In that respect the residuals will probably not be larger than what is reflected in Orbit 19, since the Schmidt potential lacks a massive halo. However, I would question whether current star counts and gas surveys can really establish that there is a non-separable model potential that is more "realistic" than a Stäckel potential in the solar neighborhood.

FLYNN: You showed some K_z laws, with limits either side due to a changing velocity ellipsoid.   Do these limits also include the effect of estimating the "catalog" from which the simulations were drawn?

STATLER: Not explicitly, but the effects of most changes in the catalog are too small to make much difference - only at the fraction of a percent level in the calculation of the likelihood.

GILMORE: I would just amplify the point you have already made.   The tilt term depends on a radial derivative and the radial derivative in the disk you are using is much steeper than in the real one.

STATLER: This is true, although I have shown how this can be remedied by using combinations of components that are more disk-like.   Also, remember that both you and I obtain about the same value for the contribution of the tilt term to the error.   Where we differ is in the shot noise contribution.   It is possible that this difference stems from your sample having a larger fraction of stars at low z than my fake test sample; clearly the best way to see if this is the case is to apply my analysis to your data.

K_z USING K DWARFS AS TRACERS

Paul L. Schechter

Massachussetts Institute of Technology

John A. R. Caldwell

South African Astronomical Observatory

K dwarfs are good tracers of the potential perpendicular to the galactic plane because they are abundant, span a narrow range in absolute magnitude at a given color and yield accurate velocities and adequate giant-dwarf separation even with low S/N spectra. We have obtained (but not yet reduced) spectra for over 400 stars spread uniformly over the range $13 < V < 17$, in the color interval $0.97 \leq V-I \leq 1.13$, which should provide velocities accurate to better than 7 km/s. These should yield estimates of the acceleration perpendicular to the galactic plane accurate to better than 10% at four distances from the plane: 250, 400, 600 and 1000 pc.

Photometry was obtained using a CCD in "time-delay-integration" mode, with the tracking on the Swope 1 m telescope turned off, and the CCD clocked at a rate close to, but intentionally different from the sidereal rate. A 20 square degree region near the SGP was scanned twice in both V and I. To augment the numbers of bright stars, another 20 square degree region was scanned reversing the leads on the telescope drive and running the telescope backwards.

The photometric data were reduced using a new point spread function fitting program, DoPHOT, designed to handle large numbers of stars and run in untended "batch" mode. Repeat scans indicate that the measurement uncertainty for a single observation using 2" pixels is 0.025 mag. A color-magnitude diagram for 15464 stars in the first 20 square degree region is given in Fig. 1. This work has been supported by the U.S. National Science Foundation through grants AST83-18504 and AST87-13888.

A. G. Davis Philip and P. K. Lu (eds.)
The Gravitational Force Perpendicular to
the Galactic Plane   143 - 145
©   1989   L. Davis Press

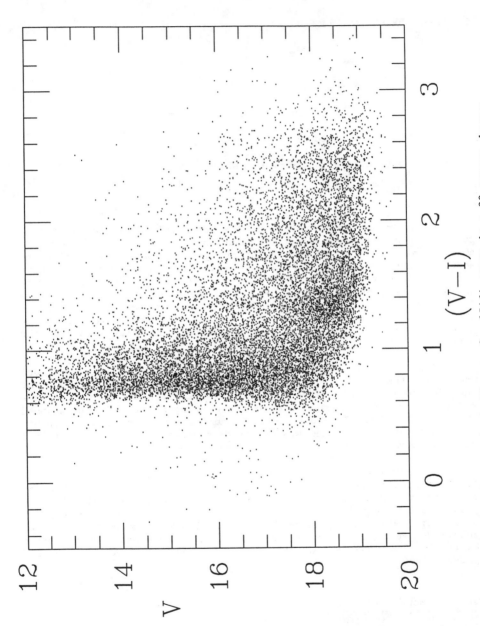

Fig. 1.   Color-Magnitude Diagram for 15464 stars in a 20 square degree region.

DISCUSSION

KING: That last graph of the giant - dwarf separation rather worries me. I wonder if you are considering going back to using MgH.

SCHECHTER: It is sheer laziness on my part that I have not used MgH. I am set up to do velocities and I get these line strengths for free, so why not use it?   Why use hot stars to study mass that is cool? Maybe it is to your advantage to use stars above this line for then you are using only cool stars to study where the mass is.   Why use a probe that is ill suited for the thing you are trying to measure.   I agree that it would be good to go back and use MgH.

CONCLUDING REMARKS

Ivan R. King

University of California, Berkeley

It is impossible to summarize a conference like this; one can only give some general impressions. (And I will also expand on one topic that I think has gotten too little attention.)

My first and strongest impression is that we are still not at all close to the answers. It seems clear that ρ_o is going to be less difficult to evaluate than Σ_o, but in neither case do I see a numerical answer in which we can really put faith. And as for K_z itself (which is after all, the mathematical symbol for the title of this conference), it seems to lie even farther in the distance.

The direction in which we have gained the most, I believe, is understanding the difficulties. Both Gould and Statler discussed the inadequacy of the present samples, particularly for the determination of Σ_o. As I recall, Gould emphasized the statistics more; I do hope that his call for a sample of 7000 stars is overly pessimistic. What Statler emphasized was the dynamical model in which any statistical discussion must be imbedded. Even if everything else is right, some one is going to have to determine how the velocity ellipsoid tilts with height above the plane.

One other difficulty, on which I would like to expand, is the choice of stellar sample. In the "old days", when our instrumentation did not allow faint stars to be reached, the traditional choices were luminous types: the main-sequence A stars and the K giants. In fact, the A stars won't work at all, because they are too young to be in the equilibrium state that the theory must assume. As for the K giants, they can be correctly used only with a much more sophisticated treatment than has ever been applied to them, a treatment that may be beyond our practical reach. The problem with the K giants is that they are such an inhomogeneous group. The "funnel effect" (to use Sandage's name for it) brings into a narrow region of the HR diagram a set of stars that have main-sequence progenitors that range from late B right on into the G's. They are of very different ages; and since velocity dispersions, and therefore scale heights, go up with age, the mixture of ages (and therefore absolute magnitudes) changes with z. I am not

A. G. Davis Philip and P. K. Lu (eds.)
The Gravitational Force Perpendicular to
the Galactic Plane   1473 - 149
©   1989   L. Davis Press

aware of anyone ever having tried to sort out the individual K giants in a K_z-sample according to absolute magnitude. To do so is, in fact, unusually difficult, because the mass differences conspire to make the spread even smaller in log g, which is the quantity that we actually measure.

Thus it appears necessary to look farther down the main sequence than the A stars. The early F's are unacceptable for the same reason as the A's. It has been recognized for 50 years that they are not uniformly distributed in the Galactic plane, with a maximum near the Sun (see, for example, Fig. 11 of McCuskey 1965); one should therefore expect them not to be in equilibrium. The late F stars are probably marginal from this point of view too. These are in fact under study by the Copenhagen group, but I have heard from them that they are able to distinguish ages and will assume dynamical equilibrium only for the stars that their photometry labels as being old enough.

Even the G stars worry me a little. Some of the oldest will have evolved upward from the main sequence, again creating a correlation between absolute magnitude and z. G stars do not reassure me as safe, unless similar precautions are taken to those in the Copenhagen studies of F stars.

Next come the main-sequence K stars, all of which must still be unevolved. I had little hope for them, on account of their faintness; but it is clear that Gilmore and Kuijken have overcome this problem. Their sample still seems to be inadequate, however. Since nearly all their stars are at large z, their data are appropriate for Σ_0 rather than ρ_0; but both Gould and Statler tell us that a much larger sample is needed. Stars of this type would also be suitable for ρ_0, if they were sampled at brighter magnitudes and therefore over a wider area of sky.

There were also encouraging results presented at this conference. Knowledge is advancing rapidly about the population mixture in the z-direction, and many observational presentations here contributed to this. Also, the methods of analysis are improving. I am particularly intrigued by the probabilistic approaches to data fitting that were presented by Gilmore and by Ratnatunga. (I use the reserved word "intrigued" only because it is impossible to appreciate a complicated approach fully from a rapid oral presentation.)

And even where I have been negative, I feel encouraged, because I for one have emerged with a much clearer view of where the difficulties lie and perhaps even of what needs to be done for them. I think that we owe a lot to Phil Lu for putting together such a stimulating and enlightening meeting.

REFERENCE

McCuskey, S. W. 1965 in _Galactic_ _Structure_, A. Blaauw and M.
    Schmidt, University of Chicago Press, Chicago, p. 1.

POSTER

PAPERS

F STARS WITHIN 400 PC TOWARDS THE NORTH GALACTIC POLE: A PROGRESS REPORT

Birgitta Nordström and Johannes Andersen

Copenhagen University Observatory and
Harvard-Smithsonian Center for Astrophysics

ABSTRACT. We report on an ongoing program to obtain radial velocities for ~1200 F dwarfs brighter than B = 11.5 and within 20° of the NGP, for which distances, metal abundances, and ages have been determined from uvbyβ photometry. These data will be combined with similar, complete and homogeneous material on a local sample of some 5000, mostly evolved F stars in an analysis of the disk kinematics. We expect the data to be completed in 1990; preliminary results indicate that proper detection of the binaries in the sample is crucial for the interpretation of such velocity data.

1. INTRODUCTION

In selecting a tracer population for investigating K_z, one wants to ensure that the samples chosen are of well-defined and homogeneous origin, and that the sample definition criteria are maintained constant for samples over a range of distances from the galactic plane. Parameters which one would like to know with good precision for all stars in the sample include distance, metal abundance, age, space motion, and duplicity. The age is of interest not only in defining a homogeneous sample, but also in judging whether the stars can be expected to have reached dynamical equilibrium. F dwarfs with accurate uvbyβ photometry, proper motions, and radial velocity data fit the above description and can be subdivided into groups of very uniform metal abundance and/or age. A sizeable fraction of them are very old. However, if velocity dispersions with small statistical sampling errors are desired for these subgroups, the total observing program becomes rather large.

This paper summarizes the plan, present status, and some preliminary results from such a coordinated program which we currently have under way in collaboration with J. Knude (Copenhagen). For more general reviews of the Copenhagen Galactic Structure programs, we refer to Strömgren (1985, 1987).

A. G. Davis Philip and P. K. Lu (eds.)
The Gravitational Force Perpendicular to
the Galactic Plane   153 - 156
©   1989   L. Davis Press

2. PLAN OF INVESTIGATION

2.1. The local sample

A major aim of our study of F stars in the direction of the North Galactic Pole is to improve the determination of K_z. This is one extension of a larger program to study the chemical and dynamical evolution of the disk(s) of our Galaxy. The first step in this project is a survey of all F stars within 50-100 pc of the Sun in order to establish a local anchor point with well-determined population frequencies and kinematics with which to compare these parameters at larger distances.

Photoelectric uvbyβ photometry of all ~15,000 F stars brighter than V = 8.3 in the entire sky was carried out by Olsen (1983). Radial velocity observations of some 5,000 (mostly evolved) stars from this sample have now been completed in collaboration with M. Mayor (Geneva), E. H. Olsen (Copenhagen), H. Lindgren (ESO), and others, using CORAVEL scanners in both hemispheres as well as the CfA echelle systems. Reduction of this large material is well under way, and the derivation of distances, ages, metal abundances, and space motions for these stars will follow over the next few months. The overall strategy of this part of the project has been discussed earlier (Andersen and Nordstrom, 1985); it is presently being followed up with CORAVEL observations of ~3,500 G dwarfs with uvby photometry in the same volume. We expect all the radial-velocity observations to be completed during 1990.

2.2. The North Galactic Pole Sample

With the radial velocity data for the local F stars in hand, we have extended the program to distances of ~400 pc in the direction of the NGP. The samples observed are drawn from a catalogue of complete uvbyβ photometry of all (~5500) A5-G0 stars within 20° of the NGP to a limit of B = 11.5, carried out by J. Knude (in preparation). With the CfA echelle system at Oak Ridge Observatory, we are conducting a program to determine radial velocities for two subsets of the sample. The first of these consists of all the (~600) Intermediate Population II (thick disk?) F dwarfs, virtually all of which should be very old. In the second sample, we observe the 541 solar-metallicity Pop. I F dwarfs used by Knude et al. (1987) to study the time evolution of the two horizontal velocity components, based on distances, proper motions, and ages only. Adding the vertical velocity component should provide information on the force experienced by these stars, as well as on the scattering mechanisms which gradually heat up the disk.

At the present time, ~1150 radial-velocity observations have been obtained of some 900 of the NGP stars (400 Pop. II and 500 Pop. I). We consider two observations separated by about a year as the minimum requirement for eliminating the more obvious variable objects, and by this criterion, 145 stars have already been completed. Closer analysis of the data will show whether this level of binary detection is adequate.

3. VELOCITY VARIABLES; BINARIES

The sample of stars for which radial velocity data are complete is still too small to yield meaningful preliminary conclusions on K_z and related subjects. However, they are already giving useful indications on the occurrence of (binary) stars with variable velocities. Obviously these must be eliminated before any detailed discussion of velocity dispersions, etc. In this connection, a comparison with the results of Sandage and Fouts (1987) may be illustrative. The region studied by us includes the NGP field in which Sandage and Fouts obtained one observation each of some 400 stars, and 76 of their stars are in common with our list (the rest of their stars are later-type and/or giant stars).

57 of these stars have now been observed two or more times by us, with average velocity errors below 1 km/s. Eight of the 57 stars (~15%) are obvious double-lined binaries or show large velocity differences. For the remaining stars, the Fouts and Sandage velocities agree well with ours in the mean, and the r.m.s. differences are consistent with their stated mean error of ~5 km/s. Among the total of 240 stars in our program which have two or more observations, about 45 or ~20% are found to have variable velocities, including ~15 double-lined binaries. It would clearly be inadvisable to attempt to derive information on population statistics or galactic kinematics from such data before the remaining stars have been reobserved for variability.

4. EXPECTED PROGRESS

We expect that the radial velocity observations of the NGP stars can be completed at the CfA in 1990 or at the latest in 1991. By then, the calibrations necessary to derive distances, abundances, ages, and space motions should be well established from our study of the nearby stars. These will include new information on the detailed chemical evolution of these stars (cf. Andersen et al., 1988). Concerning the age determinations, more definitive results should also become available on the importance of convective core overshooting in intermediate mass stars (cf. Maeder and Meynet 1989). Recent unpublished work by the writers seems to support the validity of such models, which in turn imply a considerable upwards revision of the ages of evolved F stars. The ages are expected to change by variable factors for individual stars in different parts of the HR diagram, due to the considerably different isochrone shapes of the overshooting models.

Based on the improved calibrations of the fundamental parameters, and hopefully also on improved evolutionary models, suitable tracer samples from our supply of more than 1100 candidates near the NGP can be defined and a new determination of K_z be attempted. To the extent allowed by available manpower and telescope time, we intend later to pursue similar observations of deeper, but somewhat smaller samples in the directions of both galactic poles.

ACKNOWLEDGEMENTS

The NGP radial-velocity project was planned in collaboration with Dr. J. Knude and has been supported by the Carlsberg Foundation, the Danish Natural Science Research Council, and the Smithsonian Institution. We thank Bob Davis, David Latham, Bob Stefanik, and the Oak Ridge observers for their help in obtaining and reducing the CfA data.

REFERENCES

Andersen, J., Edvardsson, B., Gustafsson, B. and Nissen, P. E. 1988 in IAU Symp. No. 132. The Impact of High S/N Spectroscopy on Stellar Physics, G. Cayrel de Strobel and M. Spite, eds., Reidel, Dordrecht, p. 441
Andersen, J. and Nordström, B. 1985 in IAU Colloq. No. 88. Stellar Radial Velocities, A. G. D. Philip and D. W. Latham, eds., L. Davis Press, Schenectady, p. 171.
Knude, J., Nielsen, H. S. and Winther, M. 1987 Astron. Astrophys. 179, 115.
Maeder, A. and Meynet, M. 1989 Astron. Astrophys. 210, 155
Olsen, E. H. 1983 Astron. Astrophys. Suppl. 54, 55.
Sandage, A. and Fouts, G. 1987 Astron. J. 93, 592.
Strömgren, B. 1985 in IAU Symp. No. 106. The Milky Way Galaxy, H. van Woerden et al., eds., Reidel, Dordrecht, p. 152.
Strömgren, B. 1987 in The Galaxy G. Gilmore and R. F. Carswell, Reidel, Dordrecht, p. 229.

THE DETERMINATION OF POPULATION TYPES OF A-TYPE STARS AT HIGH GALACTIC LATITUDES

A. G. Davis Philip

Van Vleck Observatory and Union College

ABSTRACT: Schmidt spectral surveys have been made of selected 20 square degree areas at high galactic latitudes. In some of these regions the A stars identified on the plates have been measured in the four-color system which allows a study of stellar populations to be made. In the c_1, (b-y) diagram normal Population I stars can be identified by their location close to the "reference line". The Population II stars fall above this line by about 0.2 mag or more in c_1. There is a third group of stars that fall in between these two groups. They have intermediate Δc_1, Δm_1 indices and intermediate radial velocities. The scale height of the stellar distribution of the intermediate population at the South Galactic Pole is about 1 kpc and their stellar density at 500 to 1000 pc is ~ 0.5 star per 10^6 cubic parsecs. At The North Galactic Pole the stellar density of A2 - A7 stars runs from 50 to 5 stars per 10^6 cubic parsecs for distances above the galactic plane of 200 to 500 parsecs. These characteristics, of the "intermediate" stars, match those of stars presumed to be "thick disk" stars.

1. INTRODUCTION

Some details of a Schmidt telescope high galactic latitude survey can be found in Philip (1987). Two of the regions in which the spectral classification and four-color photometry are most complete are the North and South Galactic Poles. The interstellar reddening in these regions is small ($E_{(B-V)}$ = 0 to 0.05 for the NGP and 0.02 for the SGP) which means that the intrinsic colors of the stars can be calculated with accuracy. Spectral surveys at the galactic poles have been done by many investigators. At the NGP the first survey was done by Slettebak and Stock (1949) and contained spectral classifications for stars of spectral types F2 and earlier. The spectral survey for the later type stars was published by Upgren (1963). Philip and Sanduleak (1968) and Slettebak and Brundage (1971) have published spectral catalogues of early-type stars at the South Galactic Pole. Four-color photometry has been done on the early-type stars at the poles by many investigators. Among those investigations that deal specifically with early-type stars at the NGP are Philip (1968) and

A. G. Davis Philip and P. K. Lu (eds.)
The Gravitational Force Perpendicular to
the Galactic Plane   157 - 160
©   1989   L. Davis Press

Philip and Tifft (1971) and at the SGP, Philip and Drilling (1989). In this paper the stars identified as members of an "intermediate" population will be listed and discussed.

2. THE INTERMEDIATE STARS AT THE SGP

The four-color photometry in Philip (1968) is complete in a 30 square degree region centered on the NGP and only partial in a larger area selected from the Slettebak - Stock Catalog. The four-color photometry at the SGP is complete for a larger region, 230 square degrees centered on the SGP. In Philip and Sanduleak (1968) all stars in Table II (Stars not in the CD or BD Catalogs) have been measured as well as many from Table I (Stars in the CD and BD catalogs). In this article the SGP data will be discussed and the characteristics of the "intermediate" population found there will be described. Stars were classified as members of Population I if they had δc_1 indices less than 0.07. If their δc_1 indices were between 0.07 and 0.19 they were classified as members of the intermediate group and if the δc_1 index was > 0.19 the star was classified as a Population II star. The stars classified as members of the intermediate population at the SGP are listed in Table I. The table shows the PS number (Philip and Sanduleak 1968), y mag, (b-y), c_1 and m_1, δc_1 and δm_1 (Philip and Drilling 1989). In the next column will be found distances calculated by using the relation between absolute magnitude and δc_1 determined by Crawford (1970, 1979).

$$M_v = M_{v(zams)} - f\delta c_1$$

where $\delta c_1 = c_o - c_{zams}$ and f = 10 for A-stars with (b-y) = 0.05, 9 at (b-y) = 0.10, 8 at (b-y) = 0.15 and 9 at (b-y) = 0.20. In the last column on the right the radial velocity from Rodgers (1971) is shown.

The velocity dispersion for the eight stars in Table I with radial velocities is ±33 km/s. The FHB stars had a velocity dispersion of ±99 km/s. Average δm_1 indices were calculated for A stars in the (b-y) intervals 0.10 - 0.20 and 0.20 - 0.30. The intermediate stars had average δm_1 indices of 0.014 and 0.032 compared to 0.053 and 0.046 for the FHB stars in the survey, implying a metal abundance less metal poor. The δm_1 indices for Population I stars are close to zero. If one does a rough stellar density distribution by placing the stars at their respective distances in the cone of observation and dividing by the volume of the successive sections of the cone one finds that the stellar density at distances of 600 to 1000 pc below the galactic plane is about half a star per 10^6 cubic parsecs and then drops to one tenth of a star at 1400 parsecs. Thus the scale height of the intermediate population is greater than 1 kpc.

3. PLOTS OF THE FOUR-COLOR DATA AT THE POLES

TABLE I

Intermediate Population Stars at the SGP

Name	y mag.	(b-y)	c1	m1	dc1	dm1	r	Rad V
PS 66 I	7.30	0.043	1.025	0.205	0.074	-0.002	240	---
PS 01 I	9.00	0.188	0.731	0.239	0.092	-0.052	525	---
PS 100 I	9.90	0.199	0.681	0.166	0.089	0.016	794	---
PS 102 I	10.20	0.047	1.019	0.228	0.072	-0.024	912	---
PS 45 I	10.30	0.060	1.030	0.213	0.099	-0.006	955	---
PS 67 I	10.40	0.107	0.923	0.187	0.080	0.020	1000	---
PS 55 I	10.40	0.184	0.766	0.194	0.114	-0.006	1000	---
PS 24 II	10.90	0.214	0.655	0.140	0.099	0.039	1259	52
PS 74 I	10.90	0.081	1.002	0.226	0.110	-0.019	1259	---
PS 63 I	11.00	0.136	0.883	0.197	0.093	0.007	1318	---
PS 59 I	11.20	0.097	0.935	0.199	0.075	0.009	1445	---
PS 56 I	11.30	0.163	0.877	0.156	0.191	0.040	1514	---
PS 41 II	11.60	0.214	0.719	0.154	0.163	0.025	1738	---
PS 61 II	12.10	0.017	1.054	0.128	0.070	0.066	2188	---
PS 02 II	12.24	0.066	1.053	0.174	0.130	0.032	2344	35
PS 56 II	13.00	0.173	0.787	0.138	0.107	0.064	3311	-2
PS 08 II	13.09	0.142	0.846	0.186	0.069	0.011	3467	83
PS 52 II	13.12	-0.012	1.135	0.133	0.133	0.037	3467	-34
PS 17 II	13.18	-0.001	1.176	0.116	0.171	0.063	3631	-35
PS 45 II	13.83	0.000	1.172	0.140	0.167	0.040	4898	-2
PS 42 II	14.23	0.128	0.910	0.175	0.100	0.030	5888	-14

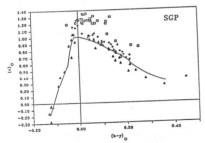

Fig. 1. c_1 index versus the (b-y) index
for FHB stars (squares), Pop. I stars
(triangles) and an Intermediate Population
(Crosses). The solid line represents the
Zero-Age Main-Sequence. South Galactic Pole.

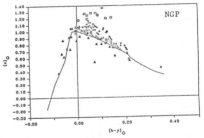

Fig. 2. Same as Fig. 1, but for stars at
the North Galactic Pole

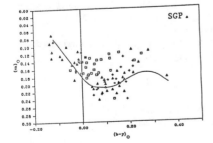

Fig. 3. m_1 index versus the (b-y) index
for FHB stars at the South Galactic Pole.
The symbols are the same as in Fig. 1.

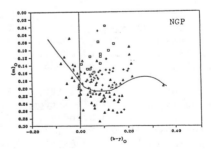

Fig. 4. Same as Fig. 3, but for stars at
the North Galactic Pole.

There are four plots comparing four-color data obtained at the SGP and NGP. Figs. 1 and 2 show dereddened c_1 versus (b-y) for the SGP and the NGP. The SGP data are from Philip and Drilling (1989) The NGP data are from Perry (1963) and Philip (1968). The line marks the position of the "reference line" or the Main Sequence. Population I stars (represented by triangles) scatter close to the reference line. The FHB stars (represented by squares) can be seen at the top of each diagram, 0.2 mag or more above the reference line. In between these two groups may be found the intermediate population (represented by crosses). Figs. 3 and 4 show the dereddened m_1 index plotted versus (b-y). Again, the triangles scatter about the reference line. The squares can be found in a distribution perpendicular to the left hand part of the reference line. The crosses scatter in between these two groups.

4.   CONCLUSIONS

In the four-color diagrams and in the Z velocity dispersion the intermediate population has characteristics that identify it as a true group with parameters in between those of Population I and Population II stars. The estimate of the scale height of 1 kpc and a metal abundance intermediate between a very low Population II metal abundance and a solar metal abundance all agree with the concept of a stellar population in a thick disk in the Milky Way Galaxy. Thus this intermediate population found at the galactic poles may be evidence in support of a thick disk.

REFERENCES

Crawford, D. L.   1970   in Stellar Rotation A. Slettebak, ed.,
     Reidel, Dordrecht, p. 114.
Crawford, D. L.   1979   in Problems of Calibration of Multicolor
     Photometric Systems, A. G. D. Philip, ed., Dudley Obs. Rept.
     No. 14, p. 23.
Perry, C. L.   1963   Thesis, Univ. of California, Berkeley.
Philip, A. G. D.   1968   Astron. J. 73, 1000.
Philip, A. G. D.   1987 in IAU Colloquium No. 95. The Second Conference
     on Faint Blue Stars, A. G. D. Philip, D. S. Hayes, and J. W.
     Liebert, eds., L. Davis Press, Schenectady, p. 509.
Philip, A. G. D. and Drilling, J. S.   1989   Astrophys J., submitted.
Philip, A. G. D. and Sanduleak, N.   1968   Bol. Ton. y Tac. 4,
     253.
Philip, A. G. D. and Tifft, L. E.   1971   Astron. J. 76, 567.
Rodgers, A. W.   1971   Astrophys. J. 165, 581
Slettebak, A. and Brundage, R. K.   1971   Astron. J. 76, 338.
Slettebak, A. and Stock, J.   1949   Astr. Abh. Hamburg No. 5.
Upgren, A. R.   1963   Astron. J. 68, 475.

STELLAR SURFACE GRAVITIES, CHEMICAL COMPOSITIONS AND RADIAL VELOCITIES FROM DIGITIZED OBJECTIVE-PRISM SPECTRA

James A. Rose

University of North Carolina

1. INTRODUCTION

A program is underway to extract quantitative information from digitized objective-prism spectra of a few tens of thousands of stars at a variety of galactic latitudes. The primary goal is to separate G dwarfs from G giants reliably and then (1) use the G dwarfs to obtain an unbiased abundance distribution for disk stars of ALL ages and (2) use the G giants to map the spatial distribution and kinematics of the intermediate, "thick disk", population.

2. OBJECTIVE-PRISM SPECTRA

Objective-prism spectra are acquired on IIa-O plates using the 10° prism on the CTIO Curtis Schmidt, yielding a dispersion of 85 Å/mm at Hδ. A 150 Å FWHM interference filter suppresses the sky background, allowing for a limiting magnitude of V ~ 13 mag in 2 hour exposures. Six exposures are obtained of each field and the digitized spectra are then coadded to increase S/N ratio. The spectra have been digitized using the Yale PDS microdensitometer.

3. SPECTRAL INDICES

Surface gravities and chemical compositions for all late-type stars (spectral types F - K) are determined from the digitized spectra using the methods described in Rose (1984, 1985a,b). The surface gravity information comes from a quantitative index that uses the gravity sensitivity of Sr II 4077 relative to Fe I lines at 4045, 4063 Å. A depression in the pseudocontinuum at 4060 Å relative to that at 4040 and 4095 Å is used to determine the metal abundance.

4. RADIAL VELOCITIES

Radial velocities for all stars are obtained using the opposed dispersion method described by Stock and Osborn (1980). For each set of six plates of each field, three are taken with the telescope east of

A. G. Davis Philip and P. K. Lu (eds.)
The Gravitational Force Perpendicular to
the Galactic Plane   161 - 164
©   1989   L. Davis Press

the pier and the other three west of the pier, which provides opposed dispersions.   This method has provided ~ 35 km/s accuracy at a dispersion of 265 Å/mm, so an accuracy of ~ 15 km/sec at a dispersion of 85 Å/mm is expected for this study.   The latter accuracy will allow for a clear kinematic separation between thick disk and standard disk.

5.   CURRENT STATUS

At this point a complete set of 6 plates has been obtained for five fields, which range from galactic latitude 45° in the galactic center direction, through the South Galactic Pole, and to galactic latitude 45° in the anticenter direction.   One set of six plates (at the SGP) has been digitized with the Yale PDS.   These spectra have been converted from density to intensity, background-subtracted, coadded, and wavelength-calibrated.   A few examples are shown in Fig. 1. Preliminary (1′ accuracy) coordinates have been derived.   Work is in progress to obtain spectral indices for all the F - K stars; R. Agostinho (Univ. of North Carolina) has begun work to obtain radial velocities.

REFERENCES.

Rose, J. A.   1984   Astron. J. **89**, 1238.
Rose, J. A.   1985a   Astron. J. **90**, 787.
Rose, J. A.   1985b   Astron. J. **90**, 803.
Stock, J. and Osborn, W.   1980   Astron. J. **86**, 246.

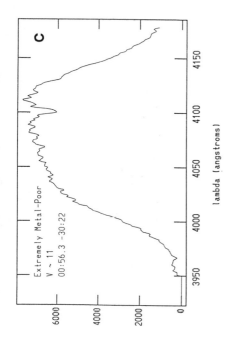

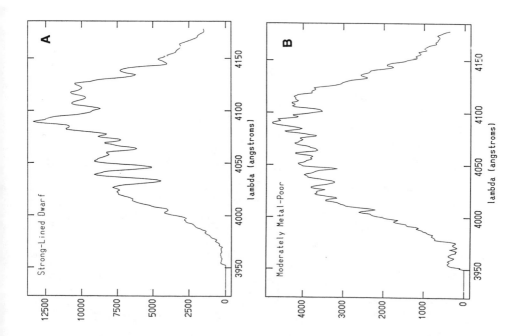

Fig. 1. Objective-prism spectra. a) Strong-lined dwarf. b) Moderately metal-poor star. c) Extremely metal-poor star.

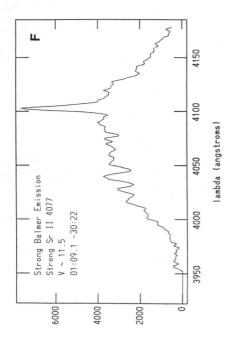

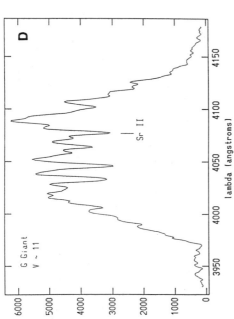

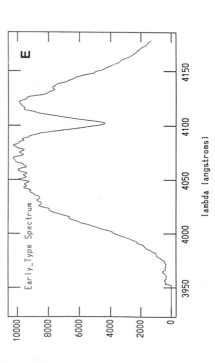

Fig. 1. Continued. d) G giant
(note strong Sr II 4077). e) Early-
type star. f) Strong Balmer emission
and strong Sr II 4077.

OBJECTIVE-PRISM SEARCH FOR RADIAL VELOCITY LIMITS

A. R. Upgren

Van Vleck Observatory

J. Stock

CIDA, Merida, Venezuela

To construct the three-dimensional distribution of any type of object within our galaxy is an extremely difficult task because of the uncertainties inherent in the determination of the distances. However, motions permit some conclusions to be made about the space distributions of the respective objects, because in a steady-state or at least nearly steady-state situation, the motions of objects must be consistent with the spatial distribution function. This is true even if we consider only one component of the motions. The determination of the radial component of the motions does not involve the knowledge of the distance, but the interpretation of the data does require at least an approximate distance for each object, if only to exclude objects beyond a limiting distance. This is necessary to insure that the sampling is restricted to a volume that is small relative to the overall dimensions of the galaxy. Spectral types and magnitudes of the precision which is obtainable from objective-prism plates are sufficient for such purposes. Peralta (1973) analyzed models of disks of unknown thickness, inclination and orientation with respect to the galactic coordinate system. His results showed that for a given direction, the radial velocities will be confined to a specific range with well defined limits and furthermore, these limits expressed as functions of galactic coordinates, are related to the above-mentioned parameters of the models. The reason for testing such models arose from the asymmetry with respect to the galactic plane, of the distributions of certain types of stars belonging to the intermediate population as found by Stock and Wroblewski (1972) and others.

It has been known for some time that radial velocities of large numbers of stars taken using the plate-pair objective prism method are most efficient in revealing these extremes in both the positive and negative values of radial velocity. A first attempt has already been made for this purpose. A survey employing objective-prism techniques was made for about 10,000 stars using the CTIO Schmidt telescope in a

A. G. Davis Philip and P. K. Lu (eds.)
The Gravitational Force Perpendicular to
the Galactic Plane   165 - 166
©   1989   L. Davis Press

mid-latitude region and shows clear upper limits to the numerical sizes of the radial velocities, positive and negative (Stock 1984).

The objective-prism method appears to still be the most efficient for this purpose. Surveys based on it can yield unbiased samples, unlike many surveys that incorporate individual spectra or photometry. They are also very efficient in the use of telescope time for their observational material. The methods used in previous work can be improved. The hand measures, even if made with a stereo measuring machine, are very time consuming. Rapid automatic measuring machines such as the PDS Microdensitometer allow the measures to be made in a short time. Weis, Upgren and Dawson (1981) have shown that the use of a PDS machine not only greatly decreases the time involved in obtaining the measures, but also increases the precision of the results. A second improvement was developed by Stock and Osborn (1980). They suggest the calculation of solutions for individual stars, once appropriate candidates are identified by means of a first general solution. A third improvement concerns the general theory of objective-prism radial velocities and the reductions used to determine them. Now it can also take advantage of the plate overlap technique, which is in common use in astrometry. In addition to the radial velocities, positions of astrometric precision are being obtained from the same measures for all objects as a byproduct.

The time-consuming hand measures involved in a project of this size delayed further work toward its completion. Now a PDS machine is being obtained by CIDA and work on it can be resumed. More than one hundred plates which had been taken with the CTIO Schmidt telescope are available for the continuation of this project. More can be taken with the same instrument and with the Schmidt telescopes at KPNO and CIDA. This will allow sampling in any direction.

REFERENCES

Peralta, J. O. 1973, Doctoral Thesis, Instituto de Fisica, Universidad Nacional de Tucuman, Tucuman, Argentina.
Stock, J. 1984 Rev. Mexicana de Astron. y Astrof. 99, 77.
Stock, J. and Osborn, W. H. 1980 Astron. J. 85, 1366.
Stock, J. and Wroblewski, H. 1972 Publ. Dep. Astron., Univ. of Chile 2, 59.
Weis, E. W., Upgren, A. R. and Dawson, D. W. 1981 Astron. J. 86, 246.

RED GIANTS NEAR THE SOUTH GALACTIC POLE: A TEST FOR AN ABUNDANCE
GRADIENT

Robert F. Wing

Ohio State University

Narrow-band photometry has been obtained for a set of M giants in
a large field at the South Galactic Pole. Since there is essentially
no interstellar reddening in this direction, abundance information can
be extracted from the measured molecular band strengths and colors.
Further, since all stars of the set are believed to be giants of
luminosity class III, their apparent magnitudes can be used as distance
indicators and a test can be made for differences in chemical
composition as a function of distance from the galactic plane.

1. THE SURVEY

The stars considered here were identified by Schiller (1981) on
blue objective-prism plates taken earlier by A. Slettebak at Cerro
Tololo Inter-American Observatory. The plates cover a field of 840
square degrees centered on the SGP and have a limiting magnitude near B
= 14.5. Schiller determined spectral types and B magnitudes for 183 M
giants and used the material to study the space density of M giants as
a function of z-distance.

2. NARROW-BAND PHOTOMETRY

The stars of Schiller's survey have been observed on the
eight-color system of narrow-band, near-infrared photometry described
by Wing (1971). This system measures the I(104) magnitude at 10400 Å,
the color temperature, and the strengths of bands of TiO and CN.
Normally, the TiO and CN indices are used to derive two-dimensional
(temperature, luminosity) spectral classifications for M stars. In
this case, however, all the program stars are considered giants since
dwarfs were excluded spectroscopically by Schiller and supergiants are
not expected in this direction. Thus we may use the CN index as an
indicator of composition. Also, the relation between TiO band strength
and color, which normally is used to determine the reddening of
individual stars, here can be used to obtain additional abundance
information: if there is a composition gradient with z-distance, we
might expect the relation between band strength and color to change as
a function of apparent magnitude.

A. G. Davis Philip and P. K. Lu (eds.)
The Gravitational Force Perpendicular to
the Galactic Plane   167 - 170
©   1989   L. Davis Press

3. BAND STRENGTH - COLOR RELATIONS

In Fig. 1 the TiO index, which measures a strong band at 7120 Å, is plotted against the reciprocal color temperature $\theta_C = 5040/T_C$, obtained from a blackbody fit to the eight-color spectrum (see White

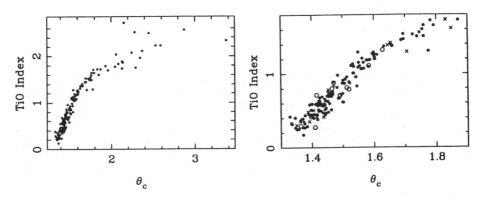

Fig. 1 (left) - Relation between TiO band strength and reciprocal color temperature $\theta_C = 5040/T_C$ for the complete Schiller sample. The relation is well defined from M0.0 (TiO = 0.25 mag) to about M5.0 (TiO = 1.4). The reddest star is the Mira RR Phe (Wing 1986).

Fig. 2 (right) - Same as Fig. 1, but plotted on an expanded scale and limited to stars earlier than M6. Crosses indicate nearby stars with I(104) < 4.0, and open circles are distant stars with I(104) > 8.0. There is no indication that the relation depends on distance.

and Wing 1978). From spectral type M0 to M5 (TiO = 0.25 to 1.40 mag) the relation is nearly linear but rather broad. For this high-latitude sample, the width of the relation cannot be blamed on interstellar reddening, nor can it be due to observational errors, which are negligible on the scale of this plot. Thus the relation must be broadened by composition differences affecting the TiO index.

Fig. 2 is an expanded version of Fig. 1, limited to stars bluer than $\theta_C = 1.9$, with different symbols distinguishing three sets of stars according to apparent I(104) magnitude: a nearby group with I(104) < 4.0, a distant group with I(104) > 8.0, and an intermediate group containing most of the survey stars. Absolute magnitudes of M giants are not well known, but for typical survey stars of types M2 - M4 we estimate that the nearby group contains stars within 400 pc and the distant group includes stars more than 2500 pc from the Sun (and from the galactic plane). We see no evidence for a change in the TiO-color relation as a function of z-distance.

In Fig. 3 the CN index is plotted against θ_C, again with stars coded according to their apparent magnitudes. Since the CN bands in M giants are quite weak, the observational errors of about ±0.01 mag in the CN index are not negligible. Nevertheless, we see here a clear systematic effect: stars with I(104) > 8.0 have systematically weak CN. On the other hand, there seems to be no systematic effect with distance for stars within about 2 kpc of the plane.

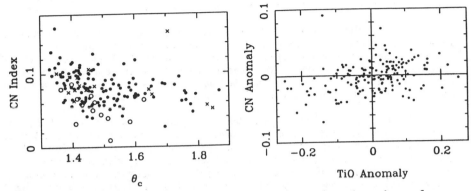

Fig. 3 (left) - Relation between CN band strength and reciprocal color temperature. Stars are distinguished by apparent I(104) magnitude as in Fig. 2; the distant group shows systematically weak CN. Note that the CN depressions in M giants seldom exceed 0.10 mag.

Fig. 4 (right) - CN anomaly vs. TiO anomaly. Stars with normal band strengths for their near-infrared colors (as defined by unreddened solar-neighborhood stars) lie near the origin. This diagram may be used to classify M giants in terms of metallicity (which affects both molecules in the same sense) and O/C ratio (which affects them in opposite senses).

4. BAND-STRENGTH ANOMALIES

Normal TiO-color and CN-color relations have been established by plotting diagrams similar to Figs. 2 and 3 for a set of nearby stars that have been classified as class III giants by P. C. Keenan. The TiO and CN "anomalies" are then defined as the differences between the observed band strengths and the normal values for stars of that value of θ_C. For this sample of unreddened class III giants, these anomalies should primarily be reflections of composition differences.

In Fig. 4, the CN anomaly is plotted against the TiO anomaly for stars with $\theta_C < 2.0$. If the anomalies are caused by general metallicity (Fe/H) differences, both molecules should be affected in the same sense and we would expect the points to be distributed along a diagonal band running from the lower-left quadrant (metal-poor stars) to the upper-right (metal-rich). Indeed, many of the stars can probably be accounted for in this way. However, the width of the band suggests that another variable is involved, one that affects the two molecules differently. This second variable is likely to be the O/C ratio, which clearly would affect TiO and CN in opposite senses. Significantly, the star in the upper-left quadrant with the largest CN excess is W Cet, the only known S star in Schiller's sample. We also note a group of stars with weak CN but strong TiO; if these stars are metal-poor, they must be even more deficient in carbon than in oxygen.

Mould and McElroy (1978) have reported that M giants in metal-rich globular clusters have weak TiO bands for their color temperatures. Unfortunately, nothing is currently known about the CN strengths of these stars. The present results would appear to indicate that high-latitude M giants, as far as 3 or 4 kpc from the plane, are not as metal-deficient as even the metal-rich globular clusters.

Since the absolute magnitudes of M giants are not well known and probably have significant dispersion, these stars are of limited value for mapping out space densities and probing gravitational force laws. However, the observations discussed here indicate that M giants may be useful in another way, in probing chemical composition differences at large distances from the plane. The sorting of M giants according to metallicity and O/C ratio in the manner proposed here will work only for unreddened samples of stars but has the advantage that it can be applied efficiently to large numbers of faint stars, yielding a new and interesting way of classifying M giants according to chemical rather than physical parameters.

ACKNOWLEDGMENTS

The photometry discussed here was obtained at Cerro Tololo Inter-American Observatory and will be published in collaboration with S. J. Schiller. I would like to thank William F. Welsh, Jr., and Pedro Saizar for preparing the figures.

REFERENCES

Mould, J. R. and McElroy, D. B. 1978 Astrophys. J., 221, 580.
Schiller, S. J. 1981 M.S. thesis, Ohio State University.
White, N. M. and Wing, R. F. 1978 Astrophys. J., 222, 209.
Wing, R. F. 1971 in Proc. Conf. on Late-Type Stars, G. W. Lockwood
    and H. M. Dyck, eds., Kitt Peak National Obs. Contr. No. 554,
    p. 145.
Wing, R. F. 1986 Bull. Amer. Astron. Soc. 18, 683.

POSTER PAPER DISCUSSION

Paper by Nordström and Andersen

GOULD: Did you say in your poster paper that the time scale for reaching tentative conclusions was next year or the year after?

NORDSTROM: I don't know if the conclusions will be ready by next year but the observations will certainly be finished within a year to a year and a half from now. We do not have a good handle on the binary population yet.

Paper by Upgren and Stock

YOSS: What is your preliminary estimate on the range of velocities?

UPGREN: Not many directions have been measured yet.

STOCK: We did a strip five degrees wide, 45 degrees long at fixed declination and when you plot the velocity as a function of right ascension there is clear evidence of a dependence on right ascension. So we want to increase the sky coverage.

Paper by Philip

NORRIS: How does your intermediate group of A stars relate to the solar abundance A stars which Rodgers has isolated far from the Galactic plane (which have a velocity dispersion ~ 60 km/s)?

PHILIP: There are three main groups of stars that I find in my high galactic latitude areas. There are the old Population II stars (characterized by a low metal abundance, large δc_1 indices and a high velocity dispersion, greater than $\pm$ 100 km/s), the Population I stars (characterized by solar metal abundance, normal c_1 indices and a low velocity dispersion) and a third group of intermediate stars (characterized by an intermediate metal-abundance, intermediate δc_1 index and intermediate velocity dispersion. The normal Population I stars seem to extend out to about 3 kpc above or below the plain in my studies. These stars, found at the NGP and the SGP, are the same group of stars later discussed by Rodgers. The intermediate group found in these studies are different because they have larger δc_1 indices.

The intermediate group does seem to be a Morgan "natural group" because in three different ways at looking at the available data, δc_1, δm_1 and Z velocity dispersion the same stars always separate out into the three groups mentioned above, the intermediate group and the Pop I and II groups.

FLYNN: What is the evolutionary status of an intermediate c_1 star?

PHILIP: I have not yet made a comparison with evolutionary models to try and find out just where these stars fit in.

Paper by Rose

RATNATUNGA: How much deeper can you go if you don't widen the spectra so much?

ROSE: If I don't widen so much (standard widening used is 0.2 mm) I can see somewhat fainter on the plate.   However, even by co-adding spectra from six plates, I don't get a sufficient S/N ratio to use these faint stars.   So I prefer to widen in order to get an increase in the S/N ratio in individual spectra.

INDICES

If a page number is underlined in the Name Index it indicates the name of an author of a paper. An underlined page number in the Object or Subject Index indicates that the object or subject was mentioned in the title of the paper. Underlined institutions are the institutions of authors of papers.

Other Publications

THE EVOLUTION OF POPULATION II STARS (1972), A. G. D. Philip, ed., Dudley Observatory Report No. 4.

MULTICOLOR PHOTOMETRY AND THE THEORETICAL HR DIAGRAM (1975), A. G. D. Philip and D. S. Hayes, eds., Dudley Observatory Report No. 9.

UBV COLOR-MAGNITUDE DIAGRAMS OF GALACTIC GLOBULAR CLUSTERS (1976), A. G. D. Philip, M. F. Cullen and R. E. White, Dudley Observatory Report No. 11.

AN ANALYSIS OF THE HAUCK-MERMILLIOD CATALOGUE OF HOMOGENEOUS FOUR-COLOR DATA (1976), A. G. D. Philip, T. M. Miller and L. J. Relyea, Dudley Observatory Report No. 12.

GALACTIC STRUCTURE IN THE DIRECTION OF THE POLAR CAPS (1977), M. F. McCarthy and A. G. D. Philip, eds., in Highlights of Astronomy, Vol 4, Reidel, Dordrecht.

IN MEMORY OF HENRY NORRIS RUSSELL (1977), A. G. D. Philip and D. H. DeVorkin, eds., Dudley Observatory Report No. 13.

IAU SYMPOSIUM NO. 80, THE HR DIAGRAM (1978), A. G. D. Philip and D. S. Hayes, eds., Reidel, Dordrecht.

IAU COLLOQUIUM NO. 47, SPECTRAL CLASSIFICATION OF THE FUTURE (1979), M. F. McCarthy, A. G. D. Philip and G. V. Coyne, eds., Vatican Observatory.

PROBLEMS OF CALIBRATION OF MULTICOLOR PHOTOMETRIC SYSTEMS (1979), A. G. D. Philip, ed., Dudley Observatory Report No. 14.

X-RAY SYMPOSIUM 1981 (1981), A. G. D. Philip, ed., L. Davis Press.

IAU COLLOQUIUM NO. 68, ASTROPHYSICAL PARAMETERS FOR GLOBULAR CLUSTERS (1981), A. G. D. Philip and D. S. Hayes, eds., L. Davis Press.

A DEEP OBJECTIVE PRISM SURVEY OF THE LARGE MAGELLANIC CLOUD FOR OB AND SUPERGIANT STARS. PART I. (1983), A. G. D. Philip and N. Sanduleak, L. Davis Press.

IAU COLLOQUIUM NO. 76, THE NEARBY STARS AND THE STELLAR LUMINOSITY FUNCTION (1983), A. G. D. Philip and A. R. Upgren, eds., L. Davis Press.

IAU SYMPOSIUM NO. 111, CALIBRATION OF FUNDAMENTAL STELLAR QUANTITIES (1985), D. S. Hayes, L. Pasinetti and A. G. D. Philip, eds., Reidel, Dordrecht.

IAU COLLOQUIUM NO. 88, STELLAR RADIAL VELOCITIES (1985), A. G. D. Philip and D. W. Latham, eds., L. Davis Press.

HORIZONTAL-BRANCH AND UV-BRIGHT STARS (1985), A. G. D. Philip, ed., L. Davis Press.

SPECTROSCOPIC AND PHOTOMETRIC CLASSIFICATION OF POPULATION II STARS (1986), A. G. D. Philip, ed., L. Davis Press.

STANDARD STARS (1986), A. G. D. Philip, ed., in Highlights of Astronomy, Vol 7.

IAU COLLOQUIUM No. 95, THE SECOND CONFERENCE ON FAINT BLUE STARS (1987), A. G. D. Philip, D. S. Hayes and J. W. Liebert, eds., L. Davis Press.

IAU SYMPOSIUM NO. 126, GLOBULAR CLUSTER SYSTEMS IN GALAXIES (1988), J. E. Grindlay and A. G. D. Philip, eds., Reidel, Dordrecht.

NEW DIRECTIONS IN SPECTROPHOTOMETRY (1988), A. G. D. Philip, D. S. Hayes and S. J. Adelman, eds., L. Davis Press.

CALIBRATION OF STELLAR AGES (1988), A. G. D. Philip, ed., L. Davis Press.

STAR CATALOGUES: A CENTENNIAL TRIBUTE TO A. N. VYSSOTSKY (1989), A. G. D. Philip and A. R. Upgren, eds., L. Davis Press.